IMPORTING THE EUROPEAN ARMY

IMPORTING THE

The Introduction of European Institutions into the

David B. Ralston

EUROPEAN ARMY

Military Techniques and Extra-European World, 1600–1914

The University of Chicago Press
Chicago & London

THE UNIVERSITY OF CHICAGO PRESS, CHICAGO 60637
THE UNIVERSITY OF CHICAGO PRESS, LTD., LONDON

Paperback edition 1996
Printed in the United States of America
99 98 97 96   5 4 3 2

Library of Congress Cataloging-in-Publication Data

Ralston, David B.
Importing the European army : the introduction of European military techniques and institutions into the extra-European world, 1600–1914 / David B. Ralston.
p. cm.
Includes bibliographical references.
ISBN 0-226-70318-5 (cloth)
ISBN 0-226-70319-3 (paper)
1. Asia—Armed Forces—History. 2. Soviet Union—Armed Forces—History. 3. Military art and science—Asia—History. 4. Military art and science—Soviet Union—History. 5. Europe—Military relations—Asia. 6. Asia—Military relations—Europe. 7. Soviet Union—Military relations—Europe. 8. Europe—Military relations—Soviet Union. 9. Military history, Modern. I. Title.
UA830.R34 1990
355′.03—dc20   89-5199
CIP

⊗ The paper used in this publication meets the minimum requirements of the American National Standard for Information Sciences—Permanence of Paper for Printed Library Materials, ANSI Z39.48-1984.

To the Memory of Gerald E. Stearn

Contents

Preface

This essay considers what happened to five societies of the world beyond the frontiers of Europe when at some point between 1600 and 1914 each undertook to reform its armed forces in accordance with current European practice. The present work is concerned less with the reform of the armed forces per se than with the ramifications of the process for the other institutions within society. Even though every one of the five countries to be studied—Russia, the Ottoman Empire, Egypt, China, and Japan—was possessed of its own unique social and cultural traits, they do have one significant quality in common. All had achieved, each in its own way and on its own terms, a noteworthy level of civilization. None was a primitive or tribal society, and in fact China might be considered, despite all of the internal problems it underwent during the nineteenth century, as the most highly civilized society in the world. Sooner or later all of them were to find themselves influenced to some degree by the dynamic civilization of Europe, and here Europe is taken to mean not only those countries which are part of that continent proper, but also those which have been founded by Europeans, especially in North America.

The introduction of the ways of Europe into the countries under discussion was not achieved by force, through outright military conquest. The thesis of this study is that it came about as a generally unforeseen, often unwelcome consequence of the development within each country of the military techniques and the modes of military organization practiced in a number of European realms. Such military reform was undertaken, consciously and voluntarily, by members of the indigenous ruling elite. This is not to argue that European influence could not result from other activities, that the ways and modes of Europe could not be introduced through trade, through missionary endeavor, or through imitation of some of its intellectual or technological achievements. But the argument here is that these channels for the transmission of European civilization were of less significance than were the efforts of a few strong-willed men in positions of power and authority to reform or reshape their armed forces along current European lines.

In four of the five cases under consideration, Europeanizing military reform was undertaken as a measure of defense against the threat posed to the political and cultural integrity of a given society by one or more vigorous, expansionist states of Europe. Only in Egypt, a more or less autonomous province of the Ottoman Empire, did the ruler not initiate military reform with a basically defensive purpose. His goal was to increase the effective independence of his realm vis-à-vis his nominal ruler, the Ottoman sultan, and even here the danger implicit in the predatory tendencies of the Europeans should not be ignored.

The process discussed in the subsequent pages has often been called *military modernization,* but the author of the present work prefers to refer to it as *the Europeanization of the armed forces.* He is not sure precisely what is meant by the term *modernization* or what it is that allows one to consider any society or social organization "modern" as opposed to one that is "nonmodern" or perhaps traditional. Nor does the author understand the criteria by which the condition of modernity is presumed to be somehow superior to its contrary condition. Although the term *modernization* will on occasion be employed in this essay, it will have a very restricted meaning. A process or an institution will have been modernized when it has been brought into reasonably close correspondence with whatever has been the current, contemporary European practice in the matter—with what is being done within the more advanced European societies, notably France, Britain, and after 1871, Germany. *Modernization* as here used is to be understood to have no other meaning or evaluative connotation than that.

That European-style military institutions can exert a powerful political and social influence within those countries outside of Europe where they have been established has been noted by scholars during the past generation. Such influence has been especially marked in societies generally considered to be "underdeveloped" or belonging to the "Third World," many of which attained independent sovereign status in the aftermath of World War II. If only because the developments in question are of relatively recent date, the scholarly work devoted to them has tended to be done by political scientists. This study has a different focus in that it purports to examine how European-style armies affected certain societies of the non-European world over a period of three hundred years or more. By that token, then, the matter is grist for the historian's mill.

Acknowledgments

In the process of writing this essay I have received assistance from a number of persons and institutions. At an early stage in the study of the subject at hand the Marion and Jasper Whiting Foundation gave me a summer grant. The Humanities Department of M.I.T. also provided research money during several summers. One term of sabbatical leave in the academic year 1977–1978 was extended to two through the good offices of Harold J. Hanham, then Dean of the School of Humanities and Social Science at M.I.T., allowing me sufficient time to get the project well launched.

For particular expertise on the countries here treated I have turned to several M.I.T. colleagues. Philip Khoury and Hasan Kayali were kind enough to read what I wrote on the Ottoman Empire and Egypt, and Peter Perdue reviewed the chapters on China and Japan. Loren Graham and Robert MacMaster did the same favor with regard to the chapter on Russia. Each of them was ready to remind me of specific historical facts and canons of interpretation.

At various stages in its incubation, the manuscript was read by Professors William McNeill of the University of Chicago, John Gagliardo of Boston University, and Shapard Clough of Columbia University, now retired, as well as by the late Gerald Stearn. I am grateful to all the aforementioned for their incisive advice and especially for their words of encouragement.

Valued colleagues among the administrative and support staff of the History Faculty of M.I.T. have contributed over the years to making a readable and eventually publishable manuscript of my typed or handwritten pages. Kathleen Bielawski, Mabel Chin, and Bernie Schlager are only the most recent collaborators in that task. The work is certainly the better for the help I have been given. Whatever may be the book's deficiencies in conception and execution, they are attributable to me.

1 *Introduction: Army, State, and Society in Europe, 1400–1700*

The irrepressible spread of European ways and European-style institutions throughout the rest of the world is probably the dominant fact in the history of the past five hundred years. By the end of the nineteenth century, Europe, or as some have characterized it, the West, had so imposed itself on most of the other societies of the world that they had been, to all intents and purposes, conquered. In the process, something like the first truly universal civilization came into being.

How Europe was able to establish its world hegemony is not something for which an explanation is necessarily self-evident. As of 1500, when this phenomenon was just getting under way, the Europeans enjoyed no significant advantages vis-à-vis other major civilizations, at least not in respect to their numbers, their technology, or their accessible natural resources. What they did seem to have was an innate restlessness, a perpetual dissatisfaction, that, despite custom and tradition, led them to look upon existing situations as "phenomena that not only could but should be changed."[1] European society had a distinctive vitality, an almost predatory vigor such that its members found themselves in constant and repeated confrontations with those of other societies. Various peoples have on occasion launched great expansive surges, rapidly extending their sway over both their neighbors and distant lands—the spread of Islam in the seventh and eighth centuries and the conquests of the Mongols are only two examples of this—but for the Europeans, expansion has been something more persistent and continuous.

What is notable about this veritable subjugation of the rest of the world is that it never involved a large number of Europeans or absorbed a significant portion of their available energies. Certain societies have shown themselves capable of mobilizing vast quantities of human and material power for the carrying out of gigantic enterprises such as the construction of the pyramids in Egypt or the Great Wall of China. These have no parallel in the West. The peculiar genius of the Europeans is manifested less in the sheer magnitude of their undertakings or the numbers of people involved than in their ability to organize their efforts so as to attain a maximum quantity of social power from a finite investment of energy.

Whatever the extent of their individual talents, Europeans have shown themselves able to think and act more effectively as members of a group than those of any other civilization. The preeminence of the Europeans in this regard seems to be founded on their capacity to conceive of and execute any large-scale endeavor, be it the administration of a business, a government, or any undertaking, as the sum of a number of discrete, interdependent operations. These operations are entrusted to persons who make the proper performance of them their paramount interest, indeed the determinant element in their social identity. In the accomplishment of the common enterprise, special considerations of technique, in particular a concern for technical efficiency, take primacy over all others, whether esthetic or moral. Decisions as to whom roles should be assigned in the common endeavor and how those so designated should be rewarded are governed by their contributions to carrying it out. Here sentiments of personal loyalty and family obligation are meant, in theory, to have no influence. One author has called this mode of managing any group undertaking *technicalism*.[2] The ability to organize groups of disparate people so that their separate energies are coordinated for the systematic, efficient achievement of a given task is similar to that rationality in social action discussed by scholars since Max Weber.

In the technicalistic society of Europe, group effort has an existence apart from the desires and aspirations of the people involved in carrying it out and is, by that token, impersonal. But if the ones so engaged can be induced to identify their own particular goals with the overall purpose of the group, its power and effectiveness are much enhanced. Other civilizations have been able to develop significant organizations which are rational by many of the same criteria. The Chinese scholar-bureaucracy and the slave army characteristic of a number of Islamic polities are two examples, but the West since ancient times has been most prolific in this regard. It was through rationally organized group endeavor that after 1500 the Europeans managed their generally successful encounters with other peoples, both in time of peace and in war.

Among the salient examples of the European propensity for the creation of rational, technicalistic organizations, and among the earliest to appear, have been the armed forces, the focus of the present study. They provide a striking demonstration of the ability of the Europeans to mobilize a multitude of diverse, individual energies and to integrate them into a single, articulated whole for the accomplishment of some larger purpose. The superiority of the West in military matters began to be evident in the course of the sixteenth century, and over the next two hundred years it only became more marked. In the armies established by the new dynastic monarchies—increasingly the predominant political conformation of the period—the principal element was an organized body of men on foot, as opposed to the hosts of mounted men in the feudal armies of the centuries just previous. The dynastic monarchs were simply rediscovering what had been appreciated by the Greeks and Romans: the effectiveness in combat of trained infantry.

Before the advent of gunpowder weapons, armed men fighting on foot either as individuals or in mass formation were highly vulnerable against men who were mounted. Only when they were drawn up as an organized tactical unit, where each supported his fellows and was in turn supported by them, did they stand a chance of holding firm, and if they were sufficiently trained or experienced, they could prevail against cavalry. A body of mounted men may have been able to keep a reasonably cohesive formation as they charged against the foe, generally cavalry like themselves. But at the moment of initial impact, they would lose their unity, their cohesion, as they became involved in a series of man-to-man duels that placed a premium on well-honed martial skills as each man tried to overcome his opponent in single combat. Thus, cavalry fighting, even if carried on by large numbers, expressed the assertive, individualistic pride characteristic of the ruling classes of medieval Europe.

The way an organized infantry force operated presupposed a much different attitude on the part of the fighting men, one stressing a sense of abnegation and even selflessness. To be a foot soldier, an ordinary person had to be conditioned through rigorous training before he and his fellows were capable of maintaining a tight, orderly formation when they moved forward into the chaos and slaughter of battle. Overcoming, even if but momentarily, their fear of death and dismemberment, and retaining presence of mind under the most shattering circumstances, they were able to act as integral, obedient members of a single, coherent body. That men could be induced to behave in that fashion may be taken as evidence of the discipline to which they were subjected. Although it has been an essential element in the development of the infantry, discipline is not a force intrinsic to or originating within a military unit by virtue of the simple fact that a body of troops has been organized. Rather, discipline must be consciously inculcated in the soldiers, its ultimate source generally lying outside the armed forces in some preexisting social or political authority, one capable of imposing it through fear if the men cannot be inspired to support its demands voluntarily.

The exemplary discipline of the armies of a number of the Greek city states and of the Roman Republic was founded on the authority of their civil institutions. It was through the disciplined skill of its infantry, the legions, that Rome grew from a little peasant republic into a world empire. In the centuries-long development of the empire, the Roman infantry was defeated in battle only infrequently, and when the legions began to decay, it was a sign and ultimately a cause of the breakdown of the imperial system of rule. Over the succeeding millennium, the fragmentation of political authority in Europe had as an almost ineluctable concomitant the eclipse of the infantry as a decisive factor in battle. The dominance of cavalry, by its nature resistant to the imposition of any real tactical discipline, may be seen as a manifestation of the inability of medieval monarchs to wield their power effectively within the lands they ruled. It is also symptomatic of a retrogression in the art of war from what it had been under the Roman Empire.

Foot soldiers may have been present on many medieval battlefields, but equipped only in rudimentary fashion and lacking both training and organization, they were held in contempt by the mounted men at arms. Even when they possessed deadly weapons, as was the case of the Genoese crossbowmen, they were still assumed to be nothing more than auxiliaries of the feudal cavalry. If, in several of the great battles of the Hundred Years' War, English archers did accomplish remarkable things at the expense of the French mounted men, the terrible losses they were able to inflict at Crécy in 1346, at Poitiers in 1356, or at Agincourt in 1415 were the result of special circumstances. The English longbowmen thus proved to be a passing phenomenon. That was not the case of their contemporaries, the Swiss pikemen. They were not a passing phenomenon. Rather, they brought about a fundamental change in the way war was fought in Europe.

The Swiss pikemen had originally been natives of the poor forest cantons of central Switzerland. Communities of free peasants, they were seeking to defend their lands against the hegemonial ambitions of their Habsburg neighbors, but they could marshal only the simplest military accoutrements in the face of their mounted, heavily armored foe. Their solution to this problem was to evolve a method of fighting whereby they were drawn up in a phalanx consisting of eight or more ranks of men, standing shoulder to shoulder, each one carrying either a long pike or a halberd. With four rows of pikes protruding from the front rank, one of them to a distance of twelve or fifteen feet, the Swiss phalanx constituted a barrier capable of breaking the most determined charge of heavy cavalry. All but invincible as a defensive formation, the pike phalanx also had offensive possibilities. A body of well-trained pikemen could move forward without breaking rank at a steady, regular pace and on contact disrupt and defeat the opposing forces, be they on foot or mounted. The efficacy of the Swiss lay not in their having evolved any new style of weaponry but rather in their reintroduction of an ordered discipline in battle. A number of special circumstances contributed to this achievement. Where the workings of the manorial and feudal systems had led to the peasants in most other parts of Europe losing much of their liberty and with it their powers of political and military initiative, the Swiss peasants were still free men with something to defend. As members of independent, self-governing communities, they were capable of organizing themselves militarily if threatened. The intense local patriotism generated within these communities led them willingly to accept the strict discipline required by their military formations.

The Swiss innovation eventually made itself felt in the major neighboring countries. Inhabitants of a poor land, lacking in fertility and natural resources, the Swiss were more than willing to export their special skills. The men of a given village or valley would emigrate as a group, in effect an already organized, cohesive tactical unit, ready to hire out to anybody wishing their services. From the start of the fifteenth until well into the sixteenth century, the armies of the emerging dynastic monarchies were likely to contain a contingent of Swiss or their imitators. It was within the framework of the new infantry formations that the first portable gunpowder weapons were introduced and made effective. Clumsy, slow-firing, and inaccurate, they would of themselves have been of little use in battle if they had not been integrated into an already existing tactical structure which compensated for their deficiencies.

The Swiss pikemen and their counterparts from other lands, in particular the German *landsknechts,* constituted a new phenomenon in Europe, one with major consequences. They were full-time soldiers having no other trade or social function, and procurable in considerable numbers. War on a large scale was coming to be waged in Europe with greater frequency. If this was due chiefly to the ambitions of the rulers of the new dynastic monarchies, political entities embodying power of a magnitude greater than anything since the days of Charlemagne or even the end of the Roman Empire, the easy availability of so many mercenary soldiers was also an important contributing element.

The armed forces of the dynastic monarchies of the early modern era were provisional in nature and offhand in organization. Usually they were raised at the start of a war or even a campaign and disbanded when it ended, with the result that "they were not tied to the state and its constitution in any lasting or systematic manner."[3] Lacking the fiscal and administrative means to maintain their forces on a regular, permanent footing, the monarchical authorities looked to military contractors to furnish troops as the need for them arose, and borrowed the sums necessary to finance the transaction from one of the great banking houses of the era. Since these all-but-independent entrepreneurs were supposed to take care of the pay and sustenance of the soldiers, a monarch generally had very little control over the armed men he had hired. Since a realm could seldom raise the sums necessary to support its forces with any degree of regularity once they were in the field, mutiny—followed by the disintegration of an army in mid-campaign—was a frequent occurrence in the sixteenth century. Few worse scourges could be let loose on a land than a mob of chronically unpaid and therefore vindictive armed men, but it was something to be expected once a large number of men had become professional soldiers with no other livelihood, when at the same time they had not yet been organized into permanent armies under the effective control of the state.

The first concerted, successful effort in Europe to establish an armed force on a more lasting basis was carried out under the Dutch Republic towards the end of the sixteenth century. Fighting for their independence from Spain in a war which was to last some eighty years, the Dutch recognized the advantages to be derived from an army maintained permanently instead of one raised in the usual spasmodic way to meet a sudden emergency. The emergency did, after all, seem to be permanent.

Because the size of the army was tailored to meet the resources of the realm, with the soldiers organized on a regular basis, they were a reliable instrument of the government. The army of the Dutch Republic reflected the prudent, rational outlook of the bourgeois ruling elite, a prosperous commercial class with no military aspirations. They were more than willing to provide the fiscal resources necessary to pay mercenaries, a majority of them foreign, rather than have their social and economic affairs continually disrupted by "arbitrary and endless demands for military manpower."[4] It was within the framework of this newly established standing army that the Stadholder of the Republic, Maurice of Orange, initiated a significant tactical innovation of which one aim was to obtain a better utilization of existing weaponry.

By the end of the sixteenth century, portable firearms had been in general use for over three generations. They may have been undergoing steady, incremental improvements, but they were still inaccurate and slow-firing. Because the men so armed were vulnerable to enemy cavalry during the long, complicated process of loading and priming their pieces, in battle they had to be protected by men carrying pikes. The tactical formation in which pikes and portable firearms were most efficaciously combined was the Spanish *tercio*, a massive square of some 3000 men. For all its formidable qualities—it was seldom bested in combat over a period of one hundred years—the *tercio* was clumsy and overly rigid, incapable of maneuvering once a battle had begun. Maurice sought to remedy these defects by breaking up the large, unwieldy formation into smaller bodies of about 550 men potentially capable of being employed with some measure of flexibility on the battlefield. The efforts were carried even further by Gustavus Adolphus, King of Sweden, whose basic tactical units consisted of 150–250 men, thereby opening new possibilities for a skilled commander. If the Maurician and Gustavian systems seemed in the eyes of contemporaries to be a striking new departure, they could also be understood as a resurrection of Roman tactics, a return to the suppleness and resilience typical of the legion in its heyday.

A necessary precondition for implementing the new tactical system was a far higher degree of training than had been required heretofore. In a *tercio* all that had been expected of a soldier was that he keep his position in rank within a massive square from which it was difficult to flee in any case. As for the subordinate officers, they really had very little to do once a battle had been engaged. A much smaller Gustavian unit, on the other hand, had to be able to maintain its tactical coherence in the heat of combat, sometimes functioning in a quasi-independent way, sometimes as part of a carefully articulated whole. Such a formation could not consist of a casual collection of soldiers assembled by a captain of mercenaries or some other military contractor on the eve of a campaign. Making unprecedented demands on the mental and moral fibre of the troops, the new infantry tactics presupposed long and constant drill as well as the inculcation of an intense discipline. Responsibility for overseeing these matters fell to a much enlarged corps of officers. Only if the armed forces were kept together on a permanent basis could both the men in the ranks and the officers develop the capabilities demanded by the new style of combat.

It was with the introduction of the Maurician and particularly the Gustavian tactical systems that the officer began to emerge as the vital element in the European army. His forerunner may have been the mounted man at arms of medieval times, but where that person had been trained exclusively to fight, an officer was henceforth supposed to do considerably more. He had to be a warrior, serving more or less on the same footing as the men under his command and sharing the perils they experienced on campaign. At the same time, an officer was endowed with a special status, keeping a certain distance from his men in order to better manage their conduct in battle. If he had to display marked qualities of leadership so that he could help to maintain their integrity as a tactical body under the most harrowing of circumstances, he was also supposed to be a selfless subordinate, who performed a variety of ordinary tasks on the order of his superiors, all the while possessing sufficient intellectual breadth to appreciate how these fitted into the larger mission of the army. It is a combination of roles easier to describe than to perform well.

In theory, an officer's rank and the likelihood of his rising within a clearly delineated hierarchy depended on his ability to carry out his assigned duties. By the mid-eighteenth century it was beginning to be recognized in a number of continental armies that the adequate fulfillment of his functions demanded more than the pragmatic knowledge acquired on the job and that some degree of preliminary schooling was also necessary. As a person trained to direct the exercise of organized violence, the officer represented a new and increasingly important professional type in European society.

For the maintenance of an army in peace as well as war a government needed financial means that were larger and procurable on a more regular basis than in the past. The haphazard fiscal methods of a medieval or early modern monarchy were clearly not adequate to the task. To raise the requisite moneys, the resources of a realm, be they material or human, had to be exploited in a sustained, systematic way. To that purpose, a number of kingdoms on the European continent developed during the course of the seventeenth and eighteenth centuries increasingly elaborate and pervasive organs of government: in a word, bureaucracies. If the army now made heavier demands on a society, the latter was at least spared the financial crises occasioned by the need to raise troops quickly at the start of hostilities, generally by means of loans obtained on exorbitant terms in a seller's market. It was also spared the social turmoil that took place when an unpaid army mutinied. From the reappearance of disciplined infantry tactics with the Swiss pike phalanx of the fourteenth century, a clear line of development may be traced, by irregular stages, down to the creation, some three hundred years later, of those impressive monuments to the capabilities for rational organization which seem to be inherent in European civilization: the permanent standing army and its concomitant, the bureaucratic state.

Military reform, along with the administrative reorganization it necessitated, often had a disruptive effect on the life of a given society. The localized feudal institutions, on which political authority had been founded and through which public life had functioned since the Middle Ages, were inexorably superseded as the dynastic monarchs sought to assert ever greater control over the potential wealth of their lands. Not only did they seek to centralize power within their kingdoms, they also endeavored, in the name of administrative efficiency, to promote greater legal uniformity among the populace as opposed to the diversity of status characteristic of medieval society. The centralizing policies of the monarchical authorities could lead to tenacious resistance on the part of those with a strong vested interest in traditional ways. On occasion, a monarch found it necessary to resort to the use of the army against his own subjects. At the very least, the inhabitants of a given realm were exposed to a new kind of social discipline, one akin to what was already current in the armed forces. The more uniform and better-disciplined a society was, the more easily it could be administered and the more predictable were the moneys it could produce, primarily for the upkeep of the armed forces.

During the seventeenth century, the France of Louis XIV was the country in Europe with the largest standing army, one capable of maintaining order at home and implementing the king's will abroad. Through the unrelenting efforts of a series of dedicated royal servants, the French monarchy managed to overcome the centrifugal forces within the realm, creating a powerful state for the support of that army. Yet for all the magnitude of this achievement, major remnants of feudalism still persisted, defying the efforts of men like Richelieu, Louvois, and Louis himself to eradicate them, and interfering with the efficient tapping of the wealth of the land for the promotion of royal policy. Even so, France by the end of the seventeenth century was able to marshal what was an unprecedented quantity of political and military power, and well she might have, since she was easily the richest, most populous country in Europe.

The rulers of lands less well endowed than France had to make more doggedly systematic efforts to mobilize available resources in order to participate in international competition with any chance of success. These realms were thus subjected to a more intense bureaucratization. In the seventeenth century, the state which had achieved the highest degree of administrative efficiency and developed the most rationally organized bureaucracy was Sweden, as in the eighteenth it was Prussia. Both countries were relatively poor, with fewer inhabitants and less readily available sources of wealth than their neighbors. Both had a series of remarkably able rulers, whose ambitions impelled their realms to assume a role on the European scene out of proportion to their real means.

At the basis of Sweden's rapid rise to great power status lay the military genius of Gustavus Adolphus as manifested in his tactical innovations and his reshaping of the country's armed forces. That at the same time the great king also undertook the reorganization of the administrative machinery of the realm may have been more than simply a coincidence. An important feature of the Gustavian reforms was the creation of modern functional departments, each responsible for a sole province of government. Further, a single system of local administration was established whereby the initiatives of the crown could be implemented uniformly in all parts of the kingdom.[5] Through her possession of such an instrument of state, Sweden was at least two generations ahead of other European countries in the efficient organization of the realm for carrying out royal policy.

To be able to wage war successfully whenever the occasion demanded, a realm endeavored not only to manage carefully its existing economic resources but also to augment them. Since increased military strength was seen as the main reason for the promotion of economic growth and development, those sectors of the economy contributing to that goal were vigorously encouraged by the state. The consequences, whether bad or good, of this mercantilist interference for an already developed and thriving economy, like that of France, may be questioned, but it seems likely that in certain poor, sparsely peopled countries, state demands led to the establishment of economic activities which could become self-sustaining and which the societies in question, if left to their own devices, would have managed to start only with great difficulty.

The conspicuous intellectual development of the era, the so-called Scientific Revolution, was significantly affected in a number of its aspects by the linked phenomena of militarism and bureaucratic growth. Science was ceasing to be a quasi-occult enterprise—the concern of a few isolated, even marginal, individuals—and was beginning to be carried on in conjunction with various institutes and academies. If in England the Royal Society was a self-run organization with its own sources of funds, in France and elsewhere on the continent its counterparts were sponsored by the government, which directed the main thrust of scientific research. The fact that a number of areas of the new science had immediate military application—metallurgy for the manufacture of guns and chemistry for the improvement of gunpowder, to mention two—provided an obvious motive for government involvement. One noteworthy result was to reinforce a particular pattern of research, namely the pursuit of scientific knowledge as something validated in terms of its utilitarian applicability rather than as knowledge for its own sake.

The evolution of education also bore the imprint of the needs of the new military bureaucratic state. In France the earliest schools devoted to the study on an advanced level of science were established to supply officers for the technical branches of the armed forces, while in Prussia and Sweden secular institutions of higher education were founded with the primary purpose of training men for positions in the state service. A bureaucratic career opened almost the only avenue for social advancement in countries not yet possessing significant manufacturing or commercial activities. By its very existence, by the demands it created and the social pressures it caused, the standing army was one of the major forces behind the almost revolutionary transformation of European society between 1500 and 1700.

Members of the ruling elite in lands around the periphery of Europe could not help but be progressively more aware of the effectiveness of European military techniques and institutions, either as they experienced it in direct confrontation or as they observed the discomfiture of others. The armed forces of these countries had heretofore been quite able to meet their military needs, but in the face of the new kind of power wielded by the states of Europe, they came to be seen as inadequate. Perceptions of that inadequacy occurred at different times and at different rates to the rulers of the countries in question. Muscovite Russia came to recognize the deficiencies in her military system soon after 1600 and during the course of the subsequent century made sporadic efforts to do something about the problem. For the Ottoman empire, although its great epic of conquest slowed down after the death of Suleyman the Magnificent in 1566, some two hundred years were to pass before people became sufficiently concerned about the declining state of the armed forces to attempt any major remedies. China and Japan, located at much greater distances from the seats of European power, would not become aware of the seriousness of the problem until the middle of the nineteenth century.

In all of these countries it occurred to some among the ruling authorities that an expeditious way to repair the evident imperfections in their armed forces might be to take something from Europe. They would introduce a number of European-style military techniques and modes of military organization. To do so, however, they had to overcome not only their own natural reluctance to tinker with time-honored ways, but also determined opposition from the more conservative and traditionally minded. Few of the would-be reformers had any inkling of how arduous an enterprise would be even the partial reshaping along European lines of the indigenous armed forces. Nor was it foreseen what the process of modernizing military reform might entail for the societies in question, or the pressures to which they would be subjected—pressures comparable in magnitude to, if not greater than, those experienced in Europe as a consequence of the development there of permanent standing armies. Indeed such military reform would seem to have been the first step in the Europeanization of those societies, a prospect not many welcomed.

2 *Military Reform under Peter the Great, His Predecessors, and His Successors*

Countries on the immediate periphery of Europe were soon made aware of the new military techniques and institutions developed by several states of that continent. Two in particular, Muscovite Russia and the Ottoman Empire, saw towards the end of the sixteenth century that they had to cope with a pressing new problem. For the Ottoman Empire, still arguably the preeminent military power in the Near and Middle East, it was a question of becoming reconciled to a slowdown in its traditional policy of expansion into southeastern Europe. Muscovy, on the other hand, was in a far more desperate situation. Over a period of some twenty-five years following the death in 1584 of Ivan IV, "The Dread," the Muscovite realm was not only wracked by internal upheavals, it was also under incessant attacks from its Western neighbors. Its very survival might well depend on its military renovation.

In the eyes of observers from the West in the late sixteenth century, Muscovy was a strange, exotic country. Muscovite Russia may have shared in many of the elements of European culture, even to the degree that many would consider her to be part of Europe, but the nature of her civilization was still quite unique. Those few who penetrated her remote, forbidding vastness and looked upon her political forms, social mores, and seemingly strange religious observances were led to wonder if Turkey were not more like Europe than was Muscovy and if indeed her inhabitants were really Christians.[1] Much of the "exotic" quality of Muscovite society was a consequence of the special circumstances of its birth and early evolution.

The loosely structured Kievan realm, forerunner to Muscovy, had maintained close ties to the Mediterranean world through trade relations with Byzantium. Once these had been shattered during the

Mongol conquest of the years 1237–1240, the Russian people were to exist in virtual isolation for some two centuries, the period in which the Muscovite autocracy took shape. Originally one of the numerous small principalities scattered across the steppe land north of the Ukraine, Muscovy owed her remarkable and steady territorial growth to the series of generally successful wars waged by her ruling princes, first against their neighbors and then, finally, against their Mongol overlords, from whose suzerainty they had liberated themselves in the course of the fifteenth century.

War has decisively influenced the evolution of most societies, but seldom has its imprint been so marked as in the case of Muscovy. The ambition of her ruling princes and the need to protect their lands against the ever-present threat of barbarian attacks from the south and east obliged them to have permanently at hand some kind of reliable armed force. Lacking any other means to remunerate their military servitors, the Muscovite rulers began to grant them tracts of land, on the yield of which they were meant to provide for themselves and to maintain their mounts and military equipment. It was during the reign of Ivan III (1462–1505) that this system of maintaining the armed forces began to be predominant.[2]

By the middle of the sixteenth century, the military servicemen had become one of the essential pillars of the evolving Muscovite state. What the servicemen needed above all was an assured supply of peasant labor, but that was in constant short supply because of the sparseness of the population, the growing availability of good land on the frontier, and the likelihood of the better-established large landholders, the so-called buyers, offering more attractive terms. The ability of the men in the service class to support themselves and to perform their military duties was thus rendered uncertain.

In response to the persistent pleas of the servicemen, the Muscovite government took a number of increasingly stringent measures to restrict peasant mobility. By the beginning of the seventeenth century the peasantry had been, to all intents and purposes, attached to the soil, although perpetual, hereditary serfdom did not become a legal fact until the enactment in 1649 of the *Ulozhenie,* or law code. A number of factors contributed to the development of serfdom, including the greed of the landholders, as well as the desire of the peasantry for security in a world of constant war and upheaval, but the main force behind it is to be found in the military needs of the Muscovite rulers.[3] So long as the peasants were deprived of the right to move about at will, they constituted a calculable, exploitable resource. They could be made to pay taxes in a regular way and to provide subsistence for the men at arms. Where in western Europe serfdom may be seen evolving in response to the generally perilous nature of existence following the decay of central authority in the late Roman and early medieval period, the imposition of serfdom in Muscovy can be understood as a manifestation of the increasing strength of the autocratic government. With the enactment of the 1649 law code, the shape of the Muscovite realm was fixed in its main lines. A vast social and political apparatus had been brought into existence primarily to promote the military ambitions of the ruler. All of the inhabitants of Muscovy were bound in some way, direct or indirect, to the carrying out of those ambitions.

One of the notable features of the populace over whom the Muscovite princes, or *tsars* as they came to be called, ruled was its intense religiosity. The Greek Orthodox faith had come to the Russian people through the conversion of Prince Vladimir in 988 and was not extirpated during the period of Mongol domination. Eventually it served to inspire them in their centuries-long struggle for liberation from the Mongol or Tatar infidels. It was as Christians that the Russian people defined themselves collectively; witness the prevailing word for peasant, *krestianin*. The leadership of the princes of Moscow in the holy fight was bolstered by the religious aura surrounding the city following the removal there in 1326 of the seat of the Metropolitan of the Russian church. One author has characterized Muscovy in this era as resembling "an expectant revivalist camp."[4] The role of religion in the development of Muscovy is of an importance comparable to the ambitions of the ruling princes.

Despite their pretensions as autocrats, the tsars found it difficult to wield effectively more than a fraction of the power they supposedly possessed. They may have sought to rule over a vast land as absolute monarchs, but they had at their disposal rather imperfect instruments of government. As the tasks to be discharged by the autocracy had grown, more or less permanent administrative departments, known as *prikazy*, had been formed in an ad hoc fashion. Usually established to meet some immediate pressing need and then never abolished, they were capricious in their powers and uncertain in their jurisdiction, often functioning in outright opposition one to another. This system has been summarized as "concentration without coordination."[5] There was no corps of trained or competent personnel capable of making the clumsy apparatus of government function with a measure of efficiency, nor did there exist any instruments through which the tsar might exercise some kind of regular supervision over its operations.

Despite the various impediments to the exercise of the absolute power that they claimed, the accomplishments of the Muscovite rulers were impressive. Their relentless, driving determination had led to the creation of armed forces capable of steadily enlarging their domain until it comprised most of present-day European Russia and stretched eastward some hundreds of miles beyond the Urals. Further, the ruling princes had evolved a crude, but workable system to support their armed forces as well as a centralized administration capable of locating the widely dispersed populace when its support was needed.[6]

From the time of Ivan I in the early part of the fourteenth century down to the reign of Ivan IV some 250 years later, the chief external danger to Muscovy was seen as coming from the Mongols or Tatars. The military system developed to cope with the Tatar threat was organized around mounted men armed with bows and arrows. Within these forces there was no division of labor nor pretension to special expertise. Moreover, among the cavalry archers there was constant bickering over place and precedence.[7] Lacking effective discipline and a clear chain of command, the Muscovite forces were usually drawn up for battle in five loosely organized masses. After they had shot off a volley of arrows, they would then charge headlong against the enemy with sabers swinging to engage them in hand-to-hand combat, while behind them would come a mob of men on foot. These rough and ready tactics were effective enough against the Tatars, who fought in essentially the same way.[8]

The Muscovite men of service were not organized as a permanent or standing military force. Rather they mustered only in the event of war or for periodic tours of duty guarding certain regions against Tatar raids. Then one half would serve during the first part of the campaigning season, the spring and early summer, while the rest of the contingent would mount guard into the fall. Other than that, the military duties of the Muscovite cavalrymen were not onerous—in theory, an annual review to check their military preparedness and the condition of their equipment. There were no regular peacetime training sessions. Although they might receive a small cash payment from the government at the start of a campaign, in general they were supposed to rely on their own estates for their maintenance and that of their mounts.[9] These militialike forces were incapable of withstanding the rigors of a long campaign and tended to disintegrate of their own accord after a period of time, but so long as the Tatars were the paramount foe, they were quite adequate for the defense of Muscovite territory. They were to prove less effective in the face of the new danger emerging on the western front.

Since the late Middle Ages, Muscovy had periodically fought the Lithuanians, the Poles, and the Swedes, as well as the Teutonic Knights. Compared to the Tatars, these foes were for a long time a cause of lesser concern, but in the course of the sixteenth century, they would come to pose an ever more insistent danger. During the Time of Troubles, between 1604 and 1613, the country suffered ruinous invasions at the hands of the Swedes and particularly the Poles, who twice captured Moscow. However much the unsettled political and social conditions of the times may have contributed to Muscovy's discomfiture in the face of these invasions, the chief problem arose from the inadequacy of her armed forces and the evident inferiority of her methods of waging war relative to those of her adversaries from the West.[10]

To meet the new conditions of warfare, the government had to procure artillery in increasing quantities, while as early as 1512 it established its first regular infantry formations, some 1,000 men equipped with matchlocks. Then in 1550 a corps of about 3,000 musketeers, or *streltsy,* was organized. The musketeers, as opposed to the cavalry archers, were outfitted and maintained at government expense. Apparently pleased with the results of this innovation, the government increased the number of infantrymen to 20,000 or more by the end of the century, although they were still a secondary part of the armed forces.[11] They continued to grow in size after that, reaching 34,000 in 1632 and 50,000 or more in 1681.[12] In battle the function of the streltsy was to deliver massed musket fire from behind protected positions in support of the main attacking force. They did not constitute the chief offensive arm as did the infantry in the West, nor can they be compared to the latter in terms of military efficacy.

Inadequate as they were proving to be against the better-organized and disciplined foes to the west, the cavalry archers continued to be the primary component of the Muscovite armed forces until the middle of the seventeenth century. The Tatars against whom they had proven their worth were still considered the chief threat by Muscovite military authorities, and not without reason. Despite the evident decline in Tatar power over the fourteenth and fifteenth centuries, the khanates that were successors to the Golden Horde were still capable of making frequent disruptive raids into Muscovite territory. They managed to sack Moscow in 1571 and, as late as 1592, penetrated the suburbs of the city.[13] When in the 1630s the country went to war against the Poles, many troops had to be drawn away from the southeastern frontier, leading to an immediate increase in the number of Tatar raids.[14] According to one scholar, the Crimean Tatars were the enemy most feared by Russian troops as late as the 1660s.[15]

Muscovy faced two different enemies on two different fronts, each of them demanding a particular kind of armed response. To create a military organization large enough and flexible enough to deal with both simultaneously was beyond the capabilities of the state as it was then constituted. It was only towards the middle of the century, when the government finally completed its ambitious system of fortifications along the southeastern frontier, combined with an extensive program of permanent settlement there, that the Tatar threat to some of Russia's more thickly populated areas was effectively surmounted.[16] By that token, the long-developing obsolescence of the cavalry archers became an accomplished fact, although they were to remain in existence for at least another three decades. Among the chief forces leading to the introduction of peasant bondage had been the needs of this military caste, and there is some irony in the final step in the imposition of serfdom, the promulgation of the 1649 law code, taking place at just about the same time that the cavalry archers became a military anachronism.

Any efforts by the government to introduce the new Western style of warfare within the framework of existing military institutions or to adapt them to its exigencies were stoutly resisted by members of the traditional military elite. No matter how obvious was the demonstrated superiority in battle of disciplined infantry formations, the typical man of service considered it socially degrading and insulting both to himself and to his family to serve in them or to accept a subordinate place in a functionally structured hierarchy of rank.[17]

Despite their pretensions to autocratic power, the recently established Romanovs did not for a long time following the Time of Troubles enjoy the kind of inherent authority possessed by rulers of the preceding dynasty, whose lineage supposedly stretched back to Rurik, the semilegendary ninth-century founder of the realm. The Romanovs had to compromise with the desires of so prominent a group in society as the mounted men of service, even though the latter had at one time been the creatures of the autocracy. Furthermore, the government lacked the means to enforce its will on the cavalry archers. It had originally granted them tracts of land for want of any other method of remuneration. Conditional as the grants were meant to be, the natural tendency among the men of service was to treat their land in proprietary fashion and to seek to have it become inheritable within their families. The throne was willing to go along with this tendency as part of the price it had to pay for their continuing support and loyalty.[18]

No more than the mounted men of service were the streltsy ready to adapt themselves to the techniques of war developed in the West. The musketeers may have been armed with portable gunpowder weapons and permanently organized into regiments which were not disbanded in time of peace, but none of this signified of itself that they were a reliable, effective military force. Mastery of the new infantry tactics presupposed a systematic discipline, a quality notably lacking in the streltsy and one which the government did not have the means or the will to enforce upon them. The state did pay the individual members of the streltsy a small stipend and gave them each a plot of land for their own maintenance, but since their pay remained about the same for a century, although the value of money was falling, the musketeers began to engage in certain trades to supplement their meager wages. Between campaigns, they gave little attention to training, preferring to concentrate on their business activities. Although these were bound to detract from the military value of the streltsy, the government did not raise serious objections, since demands on the treasury were thereby reduced.[19] As was the case with the mounted men of service, membership in the streltsy tended to become hereditary. Unlike the modern infantry formations in the west, the streltsy did not by their very existence promote any decisive changes in the existing military establishment. Rather they soon came to partake of its prevailing attitudes and practices.

So far had the military spirit of the streltsy degenerated by the middle of the seventeenth century that rank-and-file members were apt to refuse a promotion lest the increased responsibilities of a higher rank take them away from their outside economic concerns. Only about 5 percent of them participated in campaigns, in part because the government, aware of their military incapacity, had begun to assign them various internal police functions instead, but even in the maintenance of internal order, the streltsy were not a reliable body. Originally a semiautonomous corporation meant to be apart from the rest of Muscovite society, the streltsy had, through long residence among the townspeople, come to take on many of their prevailing characteristics and attitudes. In the civil disturbances which frequently punctuated Muscovy's internal history in the seventeenth century, the government could never be sure that the streltsy would not join the insurgents.[20] The streltsy may have lost most of their military utility, but they still had interests to defend and the determination to do so. With about half of them being stationed in Moscow itself, they were capable of exerting pressure on the government whenever they felt the need, and the autocracy was not willing to affront them if it could avoid doing so.

Since the government did not develop adequate fiscal resources to pay its armed forces, especially the mounted men of service, and had to resort to an essentially patrimonial system of maintaining them, it thereby sacrificed a considerable measure of control over them. When their particular methods of waging war became outmoded, the government thus lacked the leverage to compel them to change with the times. At the same time, the Romanov dynasty was also for various reasons unwilling or unable to abolish the existing military forces. The dilemma facing the autocracy tended to resolve itself, however, as the murderous nature of the new form of warfare hastened the demise of the traditional cavalry. By the 1680s their usual combat tactic of charging en masse against the foe—now men drawn up in disciplined ranks and armed with modern muskets—had so decimated their numbers that they were ceasing to be a significant force in Muscovite military and political affairs.[21] As for the streltsy, the government, recognizing their growing obsolescence, sought through successive measures to reduce their importance.

Because of the stubborn unwillingness of the old-style military forces to adapt themselves to the new Western modes of warfare, the government had to resort to the services of foreign troops in large numbers. Before each major military enterprise of the seventeenth century, agents would be sent out, mainly to the Protestant parts of Germany, to recruit mercenaries. These were expected to train and command the "troops of foreign formation," who fought according to the tactical methods then gaining currency in Europe. Foreigners made up a sizeable and increasing fraction of the Muscovite army in every war fought during the seventeenth century. In the army of 32,000 men which began the war to retake Smolensk in 1632, there were six infantry regiments commanded by foreign colonels and consisting of more than 1,500 German mercenaries and 13,000 foreign-trained Russian soldiers,[22] while it has been estimated that by 1663 27 percent of a very much larger army consisted of soldiers hired abroad.[23]

The military effort undertaken by Muscovy in the course of the seventeenth century wars against her western neighbors was on an unprecedented scale. Precise figures are difficult to come by, but it is reported that in the last stages of the victorious thirteen-year-long war of the 1660s against the Poles, the size of the army grew from

100,000 to 300,000 men. In 1681, it was given as 260,000 according to one source.[24] An armed force of that magnitude could not be raised in the casual fashion that was typical of Muscovy. In the Thirteen Years' War, some 100,000 men were drafted from among the peasantry.[25] This was the first step towards a method of recruitment which would be made systematic and permanent under Peter I.

By the end of the seventeenth century, Muscovy had made notable progress in the reorganization of her military forces along Western lines, but the government still had not established a regular standing army. For the troops of foreign formation as well as for the more traditionally organized units, the old Muscovite theory still prevailed: the state would maintain the army in time of war, but the "nation" would have to be responsible for the task in time of peace.[26] The troops were disbanded at the conclusion of hostilities and expected to provide for themselves by means of the usual agricultural pursuits. Muscovite forces in the last decades of the century were no longer essentially a militia. Rather they were, as one author has characterized them, a "semistanding army."[27] In this semistanding army, it was difficult to muster the troops for training, except intermittently, or to instill in them the kind of discipline to be found in the armies of certain European states, with predictable consequences for their military effectiveness. Still, deficient as these forces may have been by contemporary European standards, they were adequate to meet most of the military demands placed upon them.[28] The perennial problem of security against Tatar raids was resolved, while Muscovy was able to retake the land lost to Poland earlier in the century and even to advance her frontier in the Ukraine.

Apart from the impediments to military reform posed by the vested interests of the mounted men of service and the streltsy, the process was also hampered by the profound, religiously inspired xenophobia of the populace. "For a Muscovite . . . every foreigner was a dangerous infidel."[29] Standing at the center of the opposition to the introduction of Western ways, the Orthodox Church gave voice to the feeling among the masses that anything foreign should be rejected, as when in 1688 the Patriarch denounced in the bitterest terms possible the fact that foreigners held commands in the Muscovite army.[30]

Since the beginning of the sixteenth century, the ruling authorities in Muscovy had been willing to avail themselves of certain European skills and techniques, but only up to a point. No sooner did they begin to adopt something from the West than they would wonder about the wisdom of so doing and whether it would prove harmful to their faith and morals.[31] It was an attitude permeating every segment of society, even those in the ruling classes most conscious of the need to acquire European knowledge. There was thus always something hesitant about Muscovite efforts in the seventeenth century to learn from the West. Large numbers of foreigners might live in Muscovy for years and still always be objects of suspicion, residing in special segregated quarters like the German suburb of Moscow.

Muscovite intellectual life, such as it was, was under the auspices of the Orthodox Church. In this respect Muscovy may have been similar to Western Europe in the Middle Ages, but the difference lay in the fact that Russian orthodoxy, unlike Roman Catholicism, tended to understand any kind of serious intellectual endeavor as detrimental to true belief. Where in Europe it was possible for systematic secular speculation to be fostered within the framework of religious life, among the Russians any interest in these matters was looked upon as inspired by the devil and tantamount to apostasy. Military necessity more than any other factor forced the Muscovites to become seriously concerned with scientific endeavor. Evidence of the military impetus behind the development of science in Muscovy may be found in the etymology of the Russian word for *science* or *learning: nauka.* It was first introduced in a translation of a German military manual and was synonymous for "military skill." Natural science was for many years conceived of as primarily a servant of the military establishment and as having little significance apart from that function.[32] However much geometry may have anathematized as diabolical in its consequences, it did turn out to have its uses in war.

Muscovite authorities were quite aware of how effective were the new military techniques which had been developed in Europe since 1500, and they sought to introduce them into their own armed forces. To that end they made a series of efforts during the seventeenth century to organize at least a portion of the armed forces in accord with current European modes. But it should be noted that the reform effort was repeated before each major war, when the existing military establishment was hastily supplemented with large numbers of troops trained to fight in foreign formation. Although they came to constitute a majority of the forces mobilized for war, they were always disbanded at the conclusion of hostilities and had to be reconstituted whenever a new conflict loomed. Muscovy set about Westernizing her armed forces not once but on several occasions in the seventeenth century, each time starting more or less from scratch.

The implementation of Western techniques of war presupposed the existence of a standing army, something beyond the fiscal and administrative capabilities of the Muscovite realm. Marshaling the resources necessary to maintain the armed forces on a more lasting basis would have required the initiation of changes within Muscovite society too disruptive for the government to contemplate, let alone undertake. It would only be impelled to do so by the military ambitions of Peter the Great.

That the reign of Peter I marked a watershed in the development of modern Russia is a historical commonplace. During the thirty-six years he was tsar, changes amounting to a veritable revolution were initiated. Yet it would be misleading to see all of this as the outcome of a planned, coherent program of political action. As is the case with many revolutions, there was much in the final result that was accidental, while continuity with the past may be at least as striking as the radical new departures. To the degree that Peter the Great consciously pursued a single goal as ruler, it was to increase the military might of his country.

The dominant fact of Peter's reign was the war with Sweden from 1700 to 1721. It was waged for territorial gain at the expense of a neighbor, and it was in the tradition of Muscovite statecraft going back to the earliest of the ruling princes. Where it differed from prior wars was in its length and its intensity. Before that war was successfully concluded, Peter had had to bring about a major restructuring of the country's military and administrative institutions and to reorganize the bases of its ruling class, giving to the Russian state and society a physiognomy which would endure to the end of the Empire.

From his childhood Peter had a natural predilection for military matters. An exposure to these things is certainly part of the curriculum of any male member of a reigning house, but few princes would appear to have taken to them with the spontaneity and enthusiasm Peter displayed. Reared in a conventional way and exposed to conventional schooling, he was at the age of eight exiled along with his mother to the village of Preobrazhenskoe near Moscow, following a revolt by the streltsy in favor of his half-sister Sophia. From that point on, Peter's education was mostly in his own hands. "No other royal prince ever had a better claim to call himself a self-made man."[33] His chief activity in these years, aside from carousing with people from the German suburb, was the organization of two military bodies recruited first among his playmates, the so-called regiments of stableboys. Beginning in 1683, when he was eleven, Peter used his royal prerogatives to obtain the equipment necessary to build them into serious military forces. With these forces, he staged his "military games," full-scale maneuvers which required a high degree of expertise.[34] They were to evolve into the Preobrazhensky and Semonovsky Regiments of Guards, the elite nucleus of the reformed Russian armed forces.

After Peter assumed full power in 1689, having outmaneuvered his half-sister, his primary interest continued to be his military games, only now with a much enlarged scope. He displayed little concern for ordinary public affairs. In a sense, the two Azov expeditions of 1695 and 1696 were an extension of the military games, begun less in pursuit of a coherently conceived policy of state than to try out the new instruments of war he had created.[35] This first brush with the Crimean Tatars apparently whetted Peter's ambitions for further conquests in that direction, for among the reasons underlying his celebrated, precedent-breaking grand embassy to western Europe was his desire to obtain diplomatic backing for such an undertaking, as well as to learn some of the needed military and naval techniques.

While Peter was away in Europe, the streltsy rebelled, having been angered at a series of affronts, real or fancied, to their dignity and corporate interests. They also found it incomprehensible that a reigning tsar would leave the holy Russian land. Here the streltsy were acting as the self-appointed guardians of Muscovite traditionalism and as spokesmen for all those groups in the country who had been distressed by Peter's egregious behavior. The young tsar was in fact on his way home when he heard of the rebellion and simply hastened his return, arriving back in Moscow after it had already been crushed. He displayed exemplary brutality towards those responsible for the outbreak, reputedly executing a number of them himself. The remnants of the streltsy were then disbanded or dispersed. The rebellion of the streltsy provided Peter a welcome pretext to destroy a force he greatly hated.[36]

If Peter had sought to gain support in the West for some kind of Russian-led crusade against the Crimean Tatars, he was disappointed at the disinterest displayed by the European chancelleries. But while he was passing through Poland and Prussia on his way west, his attention was drawn to the inadequacy of Sweden's power for the defense of her extensive empire on the eastern and southern shores of the Baltic. Two of the northern European states, Poland and Denmark, planned to take advantage of that weakness, and although Peter apparently had given little thought to the possibility of war in the north prior to 1698, he now decided to join them. It was typical of his nonchalant approach to affairs of state that for a war which was to turn out to be the major enterprise of his reign, he devoted less than four months of preparation.[37]

From the nature of the forces raised for the war with Sweden, it would be difficult to believe that during the seventeenth century Muscovy had made considerable progress in military matters. In theory there existed a reserve of soldiers and officers trained to fight in the regiments of foreign formation, but these forces seem to have disappeared. This may have taken place because the country had not been engaged in any major conflict since 1689, so that the expensive army raised under Peter's father, Alexis, could have been considered unnecessary and thus allowed to disintegrate.[38] The army brought together to besiege the Swedish fortress of Narva on the eastern shore of the Baltic numbered about forty thousand men, mostly recent recruits under officers of foreign origin who were to prove themselves incompetent. It was put to rout by a force of some eight thousand Swedes led by their fierce and intrepid king, Charles XII, a youth of eighteen.

The defeat at Narva may be taken as marking the real start of the reign of Peter the Great. Here was revealed in the starkest possible manner the magnitude of the gap between the military capabilities of Muscovy and those of her more advanced neighbors, in particular Sweden. Poor and thinly populated she may have been, but Sweden was still a formidable adversary even if her days as a great power were coming to an end. After a momentary fit of despair, Peter threw himself with demonic energy into the task of creating an army capable of successfully meeting the forces of Charles XII. Between 1705 and 1713 about 335,000 men were raised from a population of some 15 million, a tremendous burden for the Russian people to bear.[39] Peter's recruiting methods were quite random at first, but after 1705 a system of regular conscription was introduced whereby one recruit was taken annually for every twenty taxpaying households.[40] In initiating this method to raise her armies in war and later in time of peace, Russia was almost a century in advance of the countries of western Europe, which only began to do so with the French Revolution.[41] There is considerable disagreement over the precise figures, but by the end of Peter's reign Russia had somewhere between 160,000 and 200,000 men more or less permanently under arms, or possibly as much as 1.5 percent of the total population. Since 1 percent was considered the reasonable limit in western Europe, one has a measure of the Russian achievement, especially in the light of the country's relatively backward social system and its still rudimentary organs of public administration.

Finding adequate officers to command the masses of men he had conscripted was a formidable task. In keeping with the practices of seventeenth-century Muscovy, Peter turned at first to mercenaries from abroad, since they in theory had already acquired the requisite skills; but possibly influenced by the lamentable performance of the foreign officers at Narva, he came to see the advantages of having a solid cadre of Russians in command of his troops. To obtain them, he simply conscripted men from the landed classes, as well as commissioning anyone who displayed military talent. By trial and error an increasingly large corps of capable, Russian-born officers slowly took shape.

Over the seven or eight years following Narva, Peter restored the armed forces. Carefully husbanding their strength, he built up their military qualities through minor skirmishes and small-scale engagements, all the while preparing them for more important action. "Everything was sacrificed to this end; no attention was paid to the condition of the people."[42] Of itself, the very length of the war changed the character of the armed forces. Year after year, as Peter dragged them back and forth across the Russian land, never disbanding them, they gradually acquired the attributes of a standing army without the enactment of any new law or ukase.[43] Yet one should not, perhaps, exaggerate the degree to which Peter was an innovator in promoting the emergence of this permanent regular army. He himself admitted as much in 1716 when he declared that the regularization of the armed forces had its origins in his father's activities as long ago as 1647.[44]

The need to outfit the standing army had a galvanic effect on the development of manufacturing in Russia. Here Peter was to build on an already existing but still primitive base. Since Russia possessed large resources in iron ore, which were obtainable with relative ease, a sizeable metallurgical industry producing primarily to meet the military requirements of the state had been founded some two generations before Peter came to the throne. Determined that Russia should be self-sufficient in military articles, Peter tirelessly worked to establish the necessary manufacturing facilities, either through the government itself, on its own initiative, setting up and operating the industries or by making certain designated individuals assume the task, under its close supervision. The results were impressive. It has been calculated that during Peter's reign the number of manufacturing and mining enterprises quadrupled,[45] while Russia's output of cast iron came to be among the largest in Europe.[46] In 1701, domestic production of handguns stood at a mere six thousand annually, but by 1706 it had reached thirty thousand and by 1711 some forty thousand were being produced.[47] At the time of Peter's death, Russia was capable of supplying, on the basis of her own domestic industry, all of the requirements of the army and the navy in ordnance, as well as the needs of the navy in sailcloth, rope, and lumber. The country had also gone a very long way towards meeting the demands of the army for woolen cloth from which uniforms could be made.[48]

The mercantilist policies pursued by Peter to meet the needs of the armed forces were much like those practiced in the kingdoms of western Europe. But the effect of these policies on Russian society was relatively more marked than was the case of their counterparts elsewhere. Scattered across a huge expanse of territory lacking in facilities for easy communication, the Russian people tended naturally to produce what they needed within the home. Only the state could act as a market sufficiently large to promote the growth of a real manufacturing capability, while the state also possessed the entrepreneurial drive and the resources to establish it. From Peter's time down to the present, the needs of the state, above all military, have continued to play a much larger role in the economic life of the country than has been true in western Europe.

The necessity to provide for the armed forces on a permanent basis brought about by the long war with Sweden led eventually to major changes in the Muscovite fiscal and administrative system. By the time Peter died, the country possessed at least the framework of a regularly functioning bureaucratic state on the Western model, but the initial impetus for the creation of the new structure did not come from any preconceived plan of his. In this as in so many other matters, Peter was pragmatic and empirical. He was chiefly concerned with finding ways to raise more men and money for his armed forces, and he was ready to try any expedient which seemed promising. He was also willing to drop it if it did not work.

During the first years of the struggle against Sweden, Peter was constantly on the move. Governing the realm in an ad hoc manner as he went from place to place, Peter developed the practice of sending his close associates to various parts of the country, where, as the personally designated representatives of the tsar, they were to have direct charge over obtaining the necessary men, money, and material for the war effort, thereby skirting the cumbersome system of *prikazy*.[49] This decentralized approach to administration was given definitive shape in the "provincial reform" of 1708–1711, whereby the country was divided into eight, and later ten, large districts, or *gubernii*. Each *gubernia* was responsible for the support of certain specified regiments. It was assumed that a governor on the spot appointed by the tsar would be able to get more out of the people and to get it to the armed forces in a more expeditious manner than could the *prikazy* located far away in Moscow. As a method of military maintenance in time of peace, it might have offered certain advantages, but when the war did not come to an end after the great Russian victory at Poltava in 1709 and indeed dragged on for another decade, its drawbacks became all too apparent. What Peter had created were eight separate little governments, with no single institution, no central headquarters, to coordinate their activities. Troops were habitually stationed at a great distance from the regions meant to provide for them, leading to frequent breakdowns in supply. Further, some of the old military *prikazy* remained in existence, discharging functions now assigned to the provincial governors.

Peter's first step in repairing the administrative confusion was taken in his usual offhand manner. In 1711, when he was preparing to embark upon his ill-fated campaign against the Turks on the River Pruth, Peter set up the Senate to act as his temporary surrogate while he was away, to oversee the execution and implementation of his orders. During Peter's lifetime, the Senate was not meant to have any autonomous authority, serving rather as "an administrative agent and not a political force."[50]

Although the creation of the Senate contributed greatly to the resurrection of central authority, more remained to be done. The Senate could not perform all the functions that had been the domain of the now moribund *prikazy,* and it was incapable by itself of discharging the load of work increasingly laid upon it. More elaborate administrative institutions were clearly needed. Here was one occasion in his career as a reformer when Peter seems to have set about his task with some degree of deliberation.[51] Lengthy discussions were held over the relative merits of different kinds of administrative organization, and delegates were sent to observe the practices of other countries. The collegiate organization of bureaucratic offices that existed in Sweden and Prussia apparently appealed to him more than a system whereby a single minister had responsibility for administering a given area of government. The formation of nine collegiate boards, each assigned to oversee a particular activity of state, was announced in December 1717, but it would require three more years before they were in operation. Because there were not enough native-born Russians who possessed the required degree of competence, 150 foreign experts had to be recruited to staff these new institutions of state. On paper, at least, a bureaucratic system now existed with the jurisdiction of each college defined by governmental function and covering the whole territory of the realm.

Like almost every other component of the old Muscovite state, the fiscal system was revealed to be woefully inadequate in the long war with Sweden. In his desperate search for the funds to maintain his armed forces, Peter resorted to a variety of expedients, from debasing the currency to the imposition of an almost endless series of new excise taxes. Despite his frantic efforts to increase the revenues of the state, the sums coming into the government coffers were perpetually inadequate. Indirect taxes, no matter how ingenious, simply could not provide the needed moneys.

The great fiscal innovation of Peter's reign was the poll tax. During the seventeenth century, direct taxes, primarily those exacted from individual households, had brought in about a quarter of the state revenue. A direct levy on persons had been discussed within the government even before Peter came to the throne, but the real impetus for its enactment was provided by the revelation in the census of 1710 that the number of taxable households had declined from some 800,000 in 1678, the date of the most recent census, to 640,000.[52] Many households had disappeared because of peasant flight to escape the crushing burdens imposed by the state, while hundreds of thousands of persons had perished as a result of the war or of grandiose undertakings like the construction of Saint Petersburg. A significant factor in the decline of peasant households, at least from the point of view of the government, was the tendency for two or more peasant families to come together under one roof, thus creating one official "household" and thereby lessening their tax liability. To obviate chicanery of this kind, Peter called for the levying of a single direct tax to be paid by every male above a certain age, with the exception of those in the landowning classes and the clergy.[53]

Preliminary to the imposition of the poll tax, a new census was required in order to obtain accurate figures as to the number of inhabitants in the country. Because of evasions and fraudulent reporting on the part of the populace, the government had to reject the first census conducted in 1719. Supplementary censuses were carried out in 1722 and 1723 with the help of the army, in order to overcome the resistance of the Russian people to an enumeration whose purpose they had good reason to suspect.[54] Finally a figure of 5,570,000 taxpayers, 169,000 of them town-dwellers, was arrived at and considered acceptable.

The poll tax was meant to produce revenue solely for the army, the navy being supported by funds from other sources. To arrive at the figure to be paid by each "soul," the total estimated cost of the army for a given year was divided by the number of males in the country as registered in the census. Everyone then paid exactly the same sum of money to the state with no allowances for differences in wealth or economic condition. The landowners were made responsible for collecting the poll tax from their peasants.

As far as the government was concerned, the poll tax was an immediate success. Simple in its modes of assessment and collection, it relieved the authorities of a considerable administrative burden. In its first years of collection, the new tax brought in some 2.5 or 3 million rubles more than had been produced by earlier direct taxes, thus relieving the chronically desperate state of the budget.[55] The poll tax was to remain the single largest item of revenue throughout the eighteenth century.[56] Between 1701 and 1725, government income went from 3 million to 8.5 million rubles.[57] Even if one takes into account the effects of inflation and currency debasement, the state revenues doubled during the war with Sweden, while the population, at best, did not decline. Of this revenue, army and navy took the largest portion: 80 percent in 1710, and even in a period of nominal peace at the end of the reign, two-thirds. Peter fought his wars through the unaided toil of the Russian people, who under the constant, merciless prodding of the state were able to provide the necessary money. He did not resort either to foreign subsidies or domestic borrowing. At his death, the state owed no money. The whole financial cost of the Petrine Revolution fell on one generation.[58]

At least as significant as the financial effects of the poll tax were the social consequences brought about by the conditions of its enactment. In the census preliminary to its imposition, all those liable to the tax were lumped together in a single legal category. Even after the enactment of the 1649 law code there had continued to be a degree of heterogeneity of legal status within the servile populace. This was now brought to an end. All the different categories of bonded people—be they servants or the clergy or of the laity, agricultural or household workers—were by a decree of January 19, 1723, made equally liable for the poll tax. In the Petrine state, there was now only one large class of taxpayers embracing most of the population, and it was subject to a uniform, invariable direct tax. A peasant now paid his tax to the landowner, who became in effect the financial agent and inspector of serf labor for the government. In thus implicitly delegating to the landowners such wide legal powers over their serfs, the state "abandoned them to the mercy of their masters."[59] The inauguration of the poll tax may thus be seen as a "benchmark in a long drawn-out process, the degradation of the Russian peasant."[60] Through the preparation and implementation of this single fiscal measure, undertaken primarily to finance the new, permanent standing army, there had been effected a vast simplification both in the legal structure of Russian society and in the apparatus of social power, one working to the benefit of the landowners above all. It should also be noted that it represented a relative increase in the efficiency and the power or the state, since the more uniform the legal condition of the people in a country, the less effort the state had to expend in dealing with them.

The Petrine state was a clear lineal descendant of the old Muscovite service state. In the socioeconomic system created by Peter the Great, just as in the one fathered by the Muscovite tsars of prior centuries, the peasantry at the bottom of the hierarchy paid taxes to the autocracy and worked to provide for the maintenance of its military servitors. In addition, they now had to contribute their sons to the ranks of the army on a regular basis, one annually from approximately every twenty households, to serve for twenty-five years. Where before the legal situation and the obligations of the bonded peasantry had been lacking in definition, they were now spelled out and made more stringent. Even more than the Muscovite polity before it, the bureaucratic state established by Peter I was founded on the sacrifices exacted from the peasantry.

As for those in the landowning classes, they too had sacrifices to make. Military service to the autocracy had traditionally been the duty of Muscovite landowners and indeed the condition of their being allowed to hold land, but where in the past it had not pressed too heavily, things were about to change. The easygoing days were over when the servicemen were permitted to vegetate on their estates, "making more or less casual appearances in the militia."[61] They now served on a regular basis, however unwillingly, and Peter was quite ready to coerce them if necessary. In his efforts to make his servitors perform their duties Peter had extensive and meticulous records kept of them. There were periodic inspections of their ranks such as took place in 1712, when the young men from the landowning families of Moscow were sent en masse to Saint Petersburg to be passed in review. In 1714 every man of privileged birth between the ages of twenty and thirty was required to register with the Senate under penalty of confiscation of his fortune should he not do so, while a ukase of 1720 imposed even more frightful penalties on those who failed to appear for inspection. That such orders had to be repeatedly published and often in increasingly harsh terms is suggestive of how difficult it was to make the members of the disparate and geographically scattered Russian elite fulfill their service obligations in regular fashion.[62]

At first the Senate was given the task of keeping track of them, maintaining a register of their names, assigning them to various posts, and generally following their careers. In 1722 the job was transferred to the newly created office of the *heraldmeister*.[63]

Peter was justifiably concerned about the low level of education of the Russian upper classes. The existence of a modernized army required that there be a sizable number of officers capable of commanding large, articulated bodies of men and of performing administrative functions of some degree of complexity, tasks beyond the grasp of the typical, almost illiterate Russian landowner. The tsar went to great lengths to remedy the situation, organizing an extensive system of primary schools and endeavoring to make the sons of the privileged class attend. He resorted to the most extreme and brutal threats to compel them to acquire even a smattering of literacy, along with the rudiments of mathematics, but his efforts met with only questionable success. Most of the schools founded by Peter never had a full complement of pupils, and many of them had closed by the time of his death. It would require about another generation before a significant portion of the elite came to appreciate the advantages of schooling.

Given Peter's lack of success in getting his future military servitors to attend school, a utilitarian system of on-the-job training had to suffice. An apprenticeship in the ranks of the army, or occasionally in a government office, became the normal preparation for a career in the Petrine service state. By a decree of 1714, the system was regularized in that a term of service in the ranks of the Guards was made the prerequisite for anyone seeking a commission. Thus the regiments of Guards came to take the place of colleges and universities as "training schools" in which young men of the privileged classes could prepare for their professional careers.[64]

Since there were never enough members of the old landowning elite during Peter's lifetime to man all the posts in the expanding civil and military apparatus, the tsar was willing to recruit talent wherever it could be found. He bestowed office on any capable person, even if of undistinguished birth. In so doing, he was giving witness to the predilection he had displayed since his earliest days for ability in all its guises; he was also acting in a manner compatible with the traditions of the Muscovite autocracy since at least the days of Ivan III in the fifteenth century. Like so many other aspects of the old Muscovite polity, the character of its ruling elite lacked precise definition. The aristocracy had always been an open caste, with membership being determined at different times by family origin, service to the state, or simply monarchical favor. Under Peter the aristocracy was to continue to be an open caste, at least as far as its recruitment was concerned, but its structure was to be endowed with an unwonted precision.

By a decree of January 16, 1721, every officer in the armed forces regardless of birth was declared to be a nobleman. That first step was followed a year later by the decree of January 24, 1722, instituting the Table of Ranks by which the whole structure of the nobility was subjected to a thorough-going rationalization. The decree established that in each of the three chief categories of service—court, civil administration, and armed forces—there would be fourteen ranks. Any given rank in one service would have its equivalent in terms of priority and precedence in the other two. The decree also contained a proviso to the effect that along with those commissioned in the armed forces, any person reaching the eighth rank in the other two services was considered to have obtained hereditary nobility. Thus was established a clear and systematic definition of what constituted nobility in Peter's Russia and how it could be attained. Along with rank attained by service to the military bureaucratic state went "the most valuable of all economic privileges, the right to own land worked by serf labor."[65] Even though a career in the state service was officially open to anyone with the requisite talent, the democratic aspects of the Table of Ranks should not be overemphasized. Anyone of ability might indeed enter state service, but the important posts still went far more easily to those from established families.[66]

Whatever may have been Peter's political vision at the start of his reign, by the end it had come to be that of a regulated state on the contemporary European model. Its parts were supposed to fit together in complementary and interrelated fashion, all contributing to a predetermined goal. Every person was bound to serve the state, from its first servant, the tsar himself, to the lowliest peasant, each in his own way and according to certain specific criteria. The terms of service for the peasantry were, of course, harsher than for landowners, but the latter had no free choice about whether they would fulfill their duties. Unlike the autocracy of Ivan III and his successors, which lacked a clear, defined structure, the state of Peter I did indeed have one.

Having set up a formidable apparatus of government, Peter found that it was difficult to make it work in any regular way. A system of administration adopted wholesale from lands so little similar to Muscovy could not be readily introduced there or be expected to function smoothly. The end result was that, as Klyuchevsky has pointed out, the disorder which prevailed in the country was masked by an imposing legislative facade.[67] The deficiencies in how the machinery of state functioned can be in large part explained by the apathy, corruption, and incompetence of its servitors. Accordingly Peter had to make do with other agents, usually men drawn from the army. Not only did he turn to the soldiers to obtain an acceptable census, he also entrusted them with the collection of the poll tax. They had direct responsibility for that task until 1727 and would continue to be involved in it until 1763, although now under the direction of civil officials.[68] There seemed, in fact, to be no domain of state where the military were not deemed competent according to Peter's lights. Thus in 1722 at the time he was reorganizing the Russian Church after the abolition of the Patriarchate, making it essentially a branch of the bureaucracy, his choice for the office of Procurator of the Holy Synod, the administrative head of the Church, was someone selected from among "the good and courageous army officers."[69]

In his efforts to run the country, Peter placed especial reliance on the regiment of Guards, that earliest manifestation of his restless, creative energy. The young officers of the Guards had been reared in the midst of Peter's reform efforts, and many of them were ardent supporters of what he was trying to accomplish. Peter used them to provide a motor impulse for the cumbrous machine of state, "compelling authorities high and low alike to behave themselves and to carry out the law."[70] It is indicative of the problems faced by Peter that he had to station a detachment from the Guards in the Senate to keep order among the members of that august body and to make them work.[71]

Unlike what had been Muscovite practice over the past century, Peter did not disband his army once the war with Sweden had ended. Rather, he kept it in existence, quartering the troops among the populace. Where the old Muscovite forces had been essentially a militia, and thus territorial in organization with strong local ties, those created by Peter were looked upon by the great mass of the Russian people as little more than an army of occupation, an oppressive body inflicted upon them, looming large in their everyday life and "helping to extinguish individual and local spirit."[72] For the Russian people, the reality of the Petrine regulated state turned out to be a recurrent, apparently arbitrary application of military force.

Peter sought to protect his reforms by naming his own successor, a right he claimed by a decree published in 1722. He was thereby arrogating to himself a power possessed by no prior Muscovite ruler. In tinkering so blatantly with a matter hitherto governed by tradition,

Peter was no doubt influenced by the behavior of his own son, Alexis, who never hid his antipathy for all of his father's works and who perished because of that fact. Unfortunately Peter died before he could make use of the powers he had bestowed upon himself. With no single candidate possessing an indisputable claim to the throne, two different factions comprising the leading men of the realm each tried to impose its choice; the issue was finally decided, however, when the Guards announced to the assembled dignitaries that they favored Peter's widow, Catherine. The Guards thus established their claim to a decisive voice in these matters. They would exercise it in one manner or another until the end of the century. Of Peter's immediate successors, three were women, only one of whom had the slightest dynastic claim, while of the men, one was a by of twelve, another a babe of one, and the third an idiot.[73] All of them came to the throne with an imprimatur bestowed by the Guards.

The Petrine reforms, whether they amount to a revolution or not, may be understood as having been carried out in the face of opposition from the traditional elements in the country. Members of the old service class were drafted, most of them against their will, for permanent duty in the armed forces, while representatives of the surviving boyar and magnate families saw their preeminent position in society fade before the new Petrine nobility. As for the great mass of the Russian people, they found this state to be far more systematically oppressive than any past regime. It taxed them with unprecedented thoroughness and conscripted them into the army according to a fixed and regular schedule, all the while trampling casually on their cherished customs and beliefs.

Given the general and widespread antipathy felt towards the reforms of Peter the Great, one may legitimately wonder how they survived his death. A number of factors, however, contributed to the preservation of the Petrine state. One by-product of the reforms was the demolition or neutralization of most of the institutions around which popular discontent could have been focused and effectively organized. The Church, in particular, had seen the last vestiges of its independent political and social power destroyed, as it was reduced to being simply one department of the military-bureaucratic state.

Despite the general antagonism felt by the Russian people, at least one significant group had benefitted from the Petrine Revolution. The service nobility established by Peter had a deep vested interest in the maintenance of the new sociopolitical order. For this ruling class, the primary institutions of state were not to be understood as instruments of oppression. Rather they provided the means for them to defend their power, at the same time insuring its accompanying emoluments. A strong absolutist regime, willing to use force at the slightest hint of peasant recalcitrance, was clearly the best guarantee of their privileged position, and the armed forces constituted the principal bulwark of that regime.[74] They were under legal obligation to serve the state, but whatever sense of oppression they may have felt at that fact was mitigated by its being about the only career open in the eighteenth century, one that also carried a salary. Military service had always been an essential part of the ethos of the Muscovite ruling class, even if it had been rendered in an irregular and spasmodic fashion. It continued as the chief element in the way of life of the Russian nobility after Peter the Great, only now it was placed on a more systematic basis.

During the two generations following Peter's death, the members of the nobility made persistent efforts to have alleviated some of the conditions of service he had imposed. Their efforts were generally successful. Thus in 1736 a decree was published reducing the term of service for a nobleman from life to twenty-five years,[75] while a system of rotating leaves was also established, which allowed one-third of the officers on active service at any time to return home for one year to oversee the management of their estates.[76] The climax to this determined campaign came with the publication in 1762 of the famous manifesto releasing the nobility from obligation to render service to the state.

The liberation of the Russian nobility did not result in their immediate exodus from state service. Service, which for most of them meant military service, continued to be regarded as the natural way of life for a nobleman, the key to a satisfying existence. Nor could many of them afford to foreswear the salaries they thereby received.[77] Like the Muscovite servicemen before them, most members of the eighteenth-century Russian nobility did not possess large estates capable of supporting them in ease, let alone luxury, and a salary from the state, small and irregularly paid as it may often have been, was a very welcome supplement to their revenues from the land. Also, the rulers made it clear in various manifestos that any nobleman valuing his reputation and social status would continue to serve. Only if a man had been in service could he be received at court and thus obtain access to the largesse of the ruler in terms of more land and serfs.

Military service became considerably more attractive following the end of the Great Northern War. Peter had had to contend with the constant effort on the part of the nobility somehow to escape the rigors of serving in an army constantly on campaign. Even though Russia would fight a number of wars in the eighteenth century, they would be less harrowing affairs than the long struggle against Sweden. Then too, the government in the course of the 1730s enacted a series of measures to heighten the prestige of the military and to display its particular esteem and concern for them. Service as an officer in the armed forces was thus the occupation accorded the highest respect in eighteenth-century Russia.[78]

In the course of the eighteenth century, some degree of schooling came to be accepted by the privileged class as a precondition for entry into the officer corps. Peter had founded a system of secular schools to train future officers for the armed forces, but where he had had to resort to coercion to get young men to attend, with only meager results, his successors would appear to have been more fortunate. One reason for this may be that the educational opportunities now presented were more openly and avowedly in the interests of the privileged caste, while members of the nobility rather soon understood that prior schooling was important to a successful service career and to elevation in the Table of Ranks. Within a generation of Peter's death, education was accepted as a matter of course by most of the nobility and was indeed something to be eagerly sought after.[79] When in 1731 the government established a new military academy exclusively for the sons of the nobility, the Cadet Corps, it had its full complement of students form the start, unlike the schools founded by Peter, and had to raise its quota from 200 to 360 by the time it opened its doors the next year. The Cadet Corps did present one undeniable attraction to its prospective students in that attendance there exempted future officers from having to spend a period of time in the ranks before receiving their commissions.[80] If it was by its name a military academy, the course of studies offered was wide ranging and meant to prepare students for careers in the civil service as well as the army. Over the ensuing decades a number of similar schools were established, some reserved exclusively for the sons of the nobility and a few that permitted non-nobles to enter.

At a more basic level, the army became involved in primary secular education as a system of garrison schools, each under the auspices of an individual regiment, was set up in the provinces. The regiments themselves could also function as informal institutions of learning. There the children of officers and even a few young soldiers might study the rudiments of mathematics, law, or foreign languages at the hands of a junior officer or even a noncommissioned officer. One could also acquire much information, and a broadened horizon, in the course of performing service tasks.[81]

Long exposure to regimental life had the effect of instilling in a young nobleman an essentially militaristic outlook. In the army, qualities of order, regularity, uniformity, hierarchy, and punctiliousness were held at a premium, and they could hardly help but be carried over into the civil administration as well. Although men trained specifically for a career in the civil bureaucracy were beginning to dominate in domestic administration by the end of the eighteenth century, most of them were sons of army officers and thus likely to be impregnated with a "military" way of going about things.[82] Together with the army veterans in the civil service, they helped to perpetuate in the bureaucracy of imperial Russia well into the nineteenth century a military style and manner.

Schooling and the conditions of service to the military-bureaucratic state slowly instilled in a young nobleman a new sense of social identity. The military obligations of the old Muscovite landowning elite had been in the nature of militia service. Military service of this kind had been one manifestation of a territorial identity and the consequent involvement in local affairs stemming from possession of landed property upon which they resided most of the time. Members of the new service nobility had few sustained contacts with their estates and as a result took little interest in local matters. Any young member of the landowning classes having been at age fifteen enrolled in the army and thereafter posted back and forth across the map of Russia soon ceased to identify himself with the infrequently visited family estate or the province in which it was located. His social identity was rather bound up in his regiment and his being a member of the caste of nobility devoted to state service.

Traditional Muscovy had been looked upon as the patrimony of the ruling dynasty, to be exploited and administered like a private domain. Its servitors had thought of service in terms of a personal relationship to that dynasty and how it would affect the interests of their families. The Petrine state, on the other hand, was to be understood as something apart from both the dynasty and the Russian people. A member of the nobility was no longer the servant of a particular individual, the tsar, but rather of an abstract concept, the modern power state. "For the sake of this abstraction and the glory of the Russian Empire men were organized, trained, fashioned, and ruled according to a general scheme and uniform abstract principles."[83] Already understood by the service nobility as an instrument for the protection of their interests and privileges as the new ruling class, the army also came to be in their eyes the embodiment of the ideal purposes to which their existence was devoted.

The extent of the social and psychological transformation undergone by the nobility because of the Petrine reforms may be gauged from their attitude towards matters of official rank. Status had formerly depended on one's ancestors, with genealogy determining eligibility for office. Now there was a scramble for promotion and preferment at the hands of the military-bureaucratic state, as position in the new, rational hierarchy established by Peter began to outweigh the traditional criteria based on family prominence and antiquity. "Even those born to wealth and prestige were most sensitive to questions of rank and title."[84] Mid-eighteenth-century Russia was still far from being the regulated state Peter had envisioned. Rank did not depend exclusively on demonstrated merit, while the nobility were seeking to become a closed, privileged caste. Nevertheless, a sufficiently large number of them were now devoted to the ideal of regular service for the grandiose, unwieldy apparatus of government to be able to function, albeit inefficiently, without the constant hectoring of the ruler.

Service itself acted as a catalyst for intellectual development quite apart from the acquisition of any knowledge pertaining to technical military matters. Through schooling in the Cadet Corps or through service at Saint Petersburg with the Guards, an uncouth young nobleman was exposed for the first time to the world beyond his narrow provincial circle and given the opportunity to take on a little polish as well as a smattering of European culture. He might even develop a permanent taste for literature, art, and philosophy as these were practiced in the West, thereby becoming part of the first significant group in Russia to be seriously concerned with secular culture and its implications. One author sees the numerous poems in the new neoclassical style written when "quite unexpectedly a wave of poetic enthusiasm came over the noble youths in a military school" as a significant initial manifestation of the modern Russian literature which began in the reign of Elizabeth.[85] The cultural transformation effected primarily through military service was remarkably rapid, with the result that the Russian nobleman of the 1750s, hardly one generation after the death of Peter the Great, bore very little resemblance to his grandfather or even his father.[86]

The acceptance by the nobility of the rationalistic norms of the military-bureaucratic state and the reshaping of their outlook in accordance with its demands had the effect of deepening the gulf dividing them from the great mass of the Russian people. In the past, a real element of commonality had existed between the elite and the nonelite, founded upon their similar religious beliefs and cultural assumptions. After the Petrine reforms, nobility and lower classes seemed increasingly to inhabit different worlds. Possessed of even fewer effective rights than before and preoccupied with the concrete issues of food and shelter, the peasantry held if anything more fervently to their traditional ways and communal values against the callous incursions of the new secular state. The nobility were now the dedicated servants of that accursed institution, and promotion within its ranks had become their chief concern. Awakened to an interest in various aspects of Western culture, they now saw themselves as having little in common with the rude, benighted peasantry.

The growing divorce between the elite and the nonelite could be seen in terms of externals. In Muscovy, the ruling classes may have been more splendidly arrayed than the lower orders, but the general style and cut of the clothing worn by both groups were similar, as were also their bearded faces. Now the nobility were clean shaven in the manner of eighteenth-century Europe and dressed in Western-style uniforms, at least during their service careers. Pictorial images of all the tsars down to Peter I have them attired in rich Muscovite robes; after Peter they are generally depicted in uniform, bespeaking their overriding interest in military matters. Peter III and Paul were fanatics in this regard, focusing particularly on the minutiae of drill and uniforms, while the first three of the nineteenth-century rulers—Alexander I, Nicholas I, and Alexander II—all identified themselves closely with the well-being of the army. For them it was the fundamental institution of state. Only Nicholas II would seem to be an anomaly in this line of imperial militarists.

Even as late as the reign of Peter I, Muscovite Russia was a remote faraway land, almost terra incognita to most educated Europeans. It was also outside the workings of the European state system. By the middle of the eighteenth century, however, Russia had entered Europe with a vengeance, fully participating in the cabinet wars of the period. At one point in the Seven Years' War, her troops actually occupied Berlin and their withdrawal from the conflict at the behest of the new tsar, Peter III, that unstable admirer of Frederick the Great, may have saved Prussia from dismemberment. That so vigorous an intervention in the affairs of Europe was possible within only four decades of the death of Peter is a measure of his achievement, for Russia's population was smaller than that of France and it was scattered over an immense territory. She was still plagued by poor communications and by economic underdevelopment. What Russia did possess was an army which was comparable to any in Europe and which in more than one battle overcome the vaunted troops of the Prussian monarchy.

Important as the army may have been for the assertion of Russian interests in the affairs of Europe, the great mass of the people were conscious of it on an entirely different level. It was the one public institution whose presence touched the existence of almost every inhabitant. There was not much real specialization of administrative function until the last few decades of the eighteenth century, with the result that army chiefs occupied commanding positions in the central bureaucratic offices, while the army also intervened constantly in civil affairs.[87] Thus was perpetuated for a couple of generations at least the prominent domestic role it had been assigned under Peter the Great.

Because the maintenance of the army was the chief purpose of the civil administration of the country, the Russian people were gradually led to a new perception of the nature of authority.

> The civil hierarchy (which comprised Russian society as a whole) began to find its identity not only qualitatively in the tsar's personal power over it, but also quantitatively in the tsar's measurable need for the means to sustain the army. . . . In effect, the army hovered over the domestic regime much as the tsar image had always done, and it gradually became an attribute of the tsar image, adding the force of its concrete necessities to the traditional aura of absolute power.[88]

This new conception of authority as embodied in the army and in the Petrine state was not necessarily more congenial to the Russian people than had been the older one exemplified in the Muscovite autocracy.

In the eyes of the peasantry, the army represented a new force suddenly inserted into their lives. Annually it took away a not insignificant percentage of their number, most of them never to be seen again, all in the name of purposes they found utterly irrelevant. Where for the elite, military service constituted a widening of their horizon, providing an opportunity for exposure to new aspects of life, peasant recruits experienced it as a perpetuation, even an intensification, of the restrictions that were already their daily lot. Further, the peasantry could perceive that the military-bureaucratic state meant more regular and therefore more oppressive conditions of serfdom. It is little to be wondered at that within two generations of the death of Peter there took place the greatest of all peasant uprisings, the *Pugachevschina*.

Triggered in large part by peasant anger when the release of the nobility from the obligation of state service was not accompanied by the ending of serfdom, the Pugachev Rebellion may be seen as a tragic culmination to the Petrine reforms. Its suppression was also a grimly paradoxical tribute to the efficacy of those reforms, especially military. An army made up of peasants, but now transformed into reliable instruments of the autocracy through a harsh but rational discipline, was able to suppress with brutal effectiveness this most notable effort by the peasantry to rebel against their servitude. By so doing the army permitted the social order which had been reshaped by Peter to continue unchanged for a century. As for the rationalized, Petrine state, it lasted even longer, perhaps down to the present day in some of its aspects, never completely free of its militaristic origins.

3 *The Reform of the Ottoman Army, 1750–1914*

The Ottoman Empire was the second major extra-European realm to attempt the reshaping of its military forces along European lines. Begun quite tentatively in the middle of the eighteenth century, the process of Europeanizing military reform did not really get under way until the 1820s. It was then implemented with varying degrees of persistence and continuity until the outbreak of World War I. If the Ottoman armed forces did in the course of the nineteenth century take on many of the characteristics of contemporary European armies, they were still not able thereby to prevent the amputation of successive pieces of the territory of the Empire or its final collapse.

For a number of reasons the Ottoman ruling elements were more reluctant than their counterparts in Muscovy to initiate a far-reaching program of military reform. The perils besetting the Empire were not so immediate, nor were they so easy to define, while certain social and cultural factors got in the way. Then too, the Ottoman military past had been a singularly glorious one. It was difficult to admit deficiencies in a military system which over a period of some two hundred years, following the origins of the Ottomans as a tribal principality in central Anatolia around 1300, had led to the conquest of a vast empire. At its height in the sixteenth and seventeenth centuries, it comprised most of the present day Near East, as well as Egypt, substantial portions of North Africa, the greatest part of the Balkans, and most of the historic kingdom of Hungary. By the skill of its soldiers, by the efficacy of its governing institutions, and not least by the qualities of its first ten rulers, a series of able and intelligent men, who were "rare if not unique in the annals of dynastic succession,"[1] the Ottoman Empire was among the most durable and successful polities in the history of Islam.

Because war and conquest were fundamental conditions of the existence of the Ottoman Empire, its government was "an army before it was anything else."[2] Indeed, it would only be a slight exaggeration to characterize the whole structure of the state, and even Ottoman society itself, as auxiliary elements for the support of the armed forces. The original nucleus of the army was a cavalry force of Turkish warriors. Each of them was awarded a fief, or *timar,* on the revenues of which he was expected to provide for himself and his family. It was not to be considered his property, but so long as he rendered military service when called upon and discharged certain rudimentary administrative functions, he was allowed to hold it.[3] Like any patrimonial system of administration, it suffered from the tendency of the fief holders to place local concerns and their private interests before their obligations to their ruler. Only if there was a strong, determined authority at the center, did the timariot system function as it should. Otherwise, there was constant danger of degeneration.[4]

The one salient element in the Ottoman military forces and indeed in the whole system of government was the so-called slave army of the sultan. A feature in a number of prior Islamic empires, it was brought to a high peak of effectiveness under the Ottomans. The core of the slave army was to be found in the new troops, the *Yeni Ceri,* or as they are better known in English, the Janissaries. Originally consisting of prisoners of war and mercenaries, the Janissaries came to be recruited for the most part from youths obtained through a levy carried out periodically in the Christian portions of the Empire. The young men thus conscripted converted to Islam, after which they were placed in schools run by the state, there to be trained according to their skills and aptitudes. Those appearing to be most promising intellectually were generally destined to careers as Men of the Pen in the Ottoman administration, often achieving posts of the greatest eminence, although they were still considered to be slaves of the sultan. The majority of the young conscripts entered the fighting forces, in particular the Janissaries.[5]

In their heyday, the fifteenth and sixteenth centuries, the Janissaries were incomparable soldiers, strictly trained, admirably disciplined, and inculcated with the ideal of absolute devotion to the sultan. They did not constitute the largest element in the army, but they generally were the ones who decided the outcome of a battle.[6] European observers of the era were unanimous in their favorable comments about the self-control and good order displayed by the Ottoman forces, especially the Janissaries, contrasting them with the soldiers in the mercenary armies of early modern Europe, where such qualities were difficult to find.[7] With their pay and sustenance provided by the state, the Janissaries lived as a body apart from the rest of Ottoman society. Family concerns were seen as weakening their absolute devotion to the sultan, so they were not supposed to marry until they had been pensioned off. Nor were their children, by definition freeborn Muslims and not slaves, permitted to enter the corps. Thus was precluded the creation of Janissary dynasties.

At the apogee of the Empire, entrance into the Ottoman slave army was determined primarily by ability and education.[8] This crucial sector of the ruling institution presented a remarkable spectacle, something close to a pure meritocracy. Perhaps the wonder is not that such a system of rule faltered and eventually broke down, but rather that it worked so well for so long a time. If abuses periodically appeared, they were soon suppressed at the hands of a determined sultan.

The decline of the Ottoman Empire was a long, drawn-out process, and there is by no means any general agreement among scholars concerning the relative importance of the factors involved. Nevertheless, it is clear that towards the end of the sixteenth century Ottoman institutions began to lose their vigor. Part of the explanation for the degeneration in the Ottoman system would seem to lie in developments external to the Empire. The shift in the major trade routes away from the eastern Mediterranean had unfortunate consequences for the economic life of the Empire, while the inflationary effects of the sixteenth-century price revolution were highly disruptive for an economically static, agrarian society. This economic dislocation contributed to the increasing financial difficulties besetting the Ottoman government.

Following the death of Suleyman the Magnificent in 1566, there seemed to be a weakening in the sultanal line. None of his successors possessed the vigor or the ability of the first ten rulers of the House of Osman, and several were quite remarkable for their weakness or degeneracy. Whatever the explanation for this phenomenon, it led to a relaxation in the close control the rulers had traditionally exercised over the operations of government. As a consequence, there was an intensification in the struggle for power among contending factions of the ruling class, and a lessening in the efficacy of the governing institutions of the Empire.[9] *Timars* became items of commerce to be obtained in return for nonmilitary services or for hard cash. Each *timar* still had to furnish a warrior, but the new proprietor now sent anyone he could find.[10] By the eighteenth century, the military value of the timariot system had all but disappeared.

Even more striking was the degeneration in the Janissaries. As early as the reign of Suleyman, the rule requiring Janissary celibacy had begun to be ignored, and as time went on they were recruited increasingly from among the Muslim rather than the Christian population of the Empire. Janissaries had originally been prohibited from learning or practicing an auxiliary trade while they were enrolled in the corps, but this rule too was relaxed. Soon many of them began to ignore their military duties, since these interfered with their petty craft or business interests. Most of them eventually abandoned training altogether, although they retained membership in the corps because of the pecuniary and judicial advantages connected with it. Janissary pay certificates eventually became what were in effect negotiable investments with the result that funds collected through taxation for the ostensible purpose of supporting a significant portion of the permanent armed forces of the state were "converted into a source of income for a *rentier* class."[11] If by chance they were called to duty, the Janissaries generally sent substitutes, who came to constitute a large part of the army whenever it was mobilized.[12]

However much their real military abilities may have declined, the Janissaries remained quite able to defend their own corporate interests. From an early date they were aware of their importance within the state and hence of their political power. For all their putative devotion to the sultan, they were ready to mutiny if they felt their interests were being disregarded. In the fifteenth and sixteenth centuries, however, Janissary rebellions had been only occasional phenomena; after 1600 they became regular occurrences, terrifying the inhabitants of the major cities up to and including the sultan himself.[13] As a consequence of the deterioration in the state of the Janissaries, the government found it necessary to raise new soldiers equipped with firearms and also drawing regular pay, thus adding to its already serious financial problems.

In the course of the seventeenth and eighteenth centuries, a few men in responsible positions were increasingly conscious of the Empire's decline. As far as they were concerned, the solution was obvious: all that needed to be done was to revitalize the traditional ruling institutions. Once the absolute authority of the sultan had been restored and the debilitating factional struggle within the central government ended, the men holding *timars* and the Janissaries would be made to perform their duties in the expected fashion.[14] Periodically a vigorous sultan backed by an unusually strong-willed grand vezir would attempt to do just that. Corrupt practices would be rooted out of the state administration, and the military forces of the Empire would for the moment undergo a striking revival. The achievements of the Ottoman armies during the vezirship of the successive members of the Koprulu family between 1656 and 1683, when the Empire attained its greatest territorial extent and Ottoman armies laid siege to Vienna for a second time, were examples of the most successful of these efforts at traditional reform. But then a rival faction in the ruling class would prevail, or the reforming grand vezir would be overthrown, or the sultan would lose interest. The old abuses, for a moment held in abeyance, would once again prevail.

Extending as it did over a period of centuries, the decline of the Empire, and in particular its military deterioration, were so gradual as to be of little cause for alarm to most contemporaries. Few in the ruling classes therefore felt compelled to take any serious measures to deal with the situation. Not until the Treaties of Karlowitz in 1699 and Passarowitz in 1718 with the Hasburgs did the Ottomans suffer a significant loss of territory, and even then some of it was won back within twenty years.[15] It was only in the wake of the war with Russia, lasting from 1768 to 1774, that the need to do something about the decay in the armed forces became inescapably obvious, for by the Treaty of Kucuk-Kaynarca the Ottomans had for the first time to surrender to the infidels authority over a Muslim people, the Crimean Tatars. This threatened the Ottomans' traditional self-confidence and called into question the whole rationale of a state devoted to the propagation of the faith, by force if necessary. Their sense of humiliation was heightened by the defeat having been inflicted not by one of the great European powers such as their traditional adversary, the Austrian Monarchy, but rather by a realm which until the beginning of the century had hardly been present in the consciousness of the ruling elite.

Those who began to grapple with the problem of how to reverse the decline in the fortunes of the Ottoman armed forces held it as an absolute certainty that indigenous military methods and institutions were superior to anything to be found among the infidels. This belief may have been justified in the days of Mehmed the Conqueror or Suleyman the Magnificent, but it was now becoming progressively less plausible in the light of the demonstrated effectiveness of European arms. A few were willing to admit that fact and to consider the study or even the adoption of some Western military methods. They did not find a ready audience. Prominent among those opposed to turning to the West were the religious authorities.

Religion permeated the social and political fabric of the Empire. It provided the ultimate justification for the existence of the Ottoman state, while at the same time it was intimately involved in the daily mundane concerns of its inhabitants. The religious authorities were responsible for education and expounding the law. Even the sultan, despite the absolute temporal power he possessed, found that his actions could be subject to their control. First and foremost, he was expected to rule in accordance with the *shariah,* or holy law, the interpretation of which lay in the hands of the learned men, the higher *ulema.* They claimed, and on occasion exercised, the right of deposing the sultan for failure to govern in accordance with the precepts of the *shariah.*

Within the broad spectrum of Islamic doctrine, the ulema and their followers were the upholders of Sunnism: orthodox, rigid, and legalistic. As such, they stood in opposition to the great majority of the Turkish inhabitants of the Empire, who practiced a more amorphous, popular kind of Islam, one stressing a personal approach to the deity. The more open adherents of this doctrine were to be found in the various dervish orders, especially the Bektashi. The Bektashi were the most widespread order in Anatolia and had particularly close connections with the Janissary corps, serving as chaplains to its battalions.[16] In a society where religious attitudes had profound social and political ramifications, the heterodox doctrines associated with the Bektashi placed the Janissaries in potential opposition to the ulema and ultimately even the sultan, the preeminent defender of Islamic orthodoxy.[17]

An increased stringency in the defense of Muslim values came to be characteristic of the Empire during its era of decline. In the past, when the Ottomans were on the rise and imbued with a sense of self-confidence as they pursued their holy mission of advancing the frontiers of the Domain of Islam against the infidels, they had been ready enough to look to the West and indeed to any source capable of providing them with useful techniques of war and government. They showed no hesitation about adopting the military knowledge of the Europeans or utilizing European military gadgets. But as their sense of their own power became more uncertain, especially in the face of the rapid growth in the strength and vigor of European society, the Ottomans came to see any innovation coming from the West as somehow posing a threat to the existing Islamic sociopolitical order.

Some in the ruling elite, as they contemplated the worsening performance of the Ottoman armies, were willing to react against this new orthodox defense of Muslim ways. During the course of the eighteenth century they showed themselves willing publicly to espouse the study of European military techniques. Notable among them was Ibrahim Muteferrika, a Magyar convert to Islam and founder of the earliest Turkish printing press to be established in the Ottoman Empire. The first sixteen books to be produced on it were all of a pragmatic political or military nature.[18] In a memorial he prepared for the government and had printed in 1731, Ibrahim emphasized the superiority of the military systems of Europe and advocated that they be adopted by the Ottomans. Later in the same decade, another convert to Islam, the Comte de Bonneval, formerly an officer in the armies of both Louis XIV and Prince Eugene, but now in the service of the sultan, was able to bring about certain reforms in the artillery and technical services, although Janissary opposition later nullified his efforts. A further bout of military reform was initiated during the 1770s by a French officer of Hungarian extraction, Baron de Tott, who became military advisor to the sultan. Tott's endeavors were also directed primarily towards the artillery and the engineers, and they were to prove somewhat more durable than those of his predecessor Bonneval.[19]

To the degree that both men enjoyed any success in their programs of reform, it was because their efforts were confined to the more technical branches, an area peripheral to the interests of most Ottoman soldiers. In effect, Bonneval and Tott were able to insinuate a few Western ways around the edges of the Ottoman military establishment.[20] Although both at one time or another preached the necessity of more thoroughgoing changes, it was to no avail. The opposition was too well entrenched. So, the military fortunes of the Empire continued to decline, as was demonstrated in the war against Russia and Austria which broke out in 1787. While this war was in progress, there succeeded to the throne a new sultan, Selim III.

Like many of his predecessors, Selim believed that the misfortunes besetting the Empire stemmed above all from the degenerate state into which the armed forces had been permitted to fall. In his initial approach to the problem, he broke no new ground, simply calling for a strict enforcement of the existing rules and procedures governing the Janissaries and the timariot cavalry. However, a few of his close advisers, a group recruited from among his childhood friends and slaves, did go so far as to suggest that "what had strengthened the West might also revive the East if wisely applied."[21] In other words, still more elements of contemporary European military practice should be introduced. Even these hardly radical measures could not be implemented for the moment, since Selim and his advisors had to subordinate any thoughts of reform to the necessity of making sure that the established military and religious classes would cooperate in bringing the present war to an end.[22]

Once the war was over the government began a serious attempt to restore the traditional military forces. As might have been expected, it met with little success. Officials who were sent out with full powers to inspect the rolls of all the units and to remove those not performing their service as required received no cooperation from anyone in the Janissaries or the timariot cavalry—when they were fortunate enough not to be assassinated. Further, any efforts to introduce new, up-to-date weapons within the Janissaries were stoutly resisted.[23]

In addition to trying to bring the traditional units back to their pristine state, Selim also initiated what would evolve into a major innovation, a new model army, the *Nizam-i Cedid.* A body of troops organized along European lines and armed with European-style weapons, the *Nizam-i Cedid* may have had its originating impetus less in the reforming ardor of the sultan than in a leisure-time diversion of his grand vezir. During the recent war the latter is reported to have had some Russian and Austrian prisoners of war, plus a few men from his own guard, organized and drilled in the European mode, which body of troops he then showed to the sultan. By now convinced that the Janissaries and the feudal cavalry were past redemption, Selim saw these troops as the nucleus of a new military force, one unhindered by the ways of the past. Recruiting about one hundred Turks from the streets of Istanbul and using Germans and Russians as officers and drillmasters, the grand vezir had them exercised at a spot some miles from the capital. The whole business was carried out in a clandestine way, so as not to unduly provoke the apprehension of conservatives. A separate fiscal and administrative structure was created outside of the usual machinery of state to provide the funds necessary for the initiation and subsequent maintenance of this new model army.[24]

Despite a number of organizational problems and the antagonism displayed by various conservative factions, the *Nizam-i Cedid* once established grew rapidly. Where in 1797 it consisted of one regiment of 2,536 men, by 1801 there were about 9,000 men in it and at the end of Selim's reign in 1807, some 25,000. The new formations gave a good account of themselves whenever they were used in the field either against foreign troops or against rebels.[25] Representatives from the older army units, the Janissaries in particular, were not happy over these experiments in Western military techniques. The generalized discontent felt by the conservatives was finally crystallized by Selim's purported efforts to reform the Janissary auxiliaries.[26]

In late May 1807, an uprising on the part of the sultan's conservative adversaries took place, with the Janissaries as its spearhead. In his new model army Selim would appear to have had a disciplined corps sufficiently strong to suppress any such opposition by force, but this the sultan was unwilling to do. Rather his reaction to the rebellion was to dissolve the *Nizam-i Cedid,* and in the face of this display of weakness, the Janissaries and their coinsurgents, mainly members of the ulema and students from the religious schools, were implacable. Not satisfied simply with the dissolution of the new model army, the Janissaries sought to kill as many of its soldiers as they could find.[27] A further consequence of the uprising was the promulgation by the religious authorities of a *fetva,* or legal opinion, stating that Selim's reforms were in violation of religion and tradition. They called for his deposition. Selim acquiesced in this finding and stepped down from the throne.

Selim III appears to have been fundamentally conservative in his outlook and in his vision of reform. His original intention had been to rejuvenate the Janissaries and the feudal cavalry and, eventually, to win back the lands lost through their dilapidation. Only after his efforts in that direction had failed did he seriously consider more radical measures, namely the creation of a number of European-style units. As events were to demonstrate, his commitment to this project was at best tentative.

The ease with which Selim III was overthrown is indicative of how narrow was his base of support. In their hostility to Europeanizing military reform, the conservatives within the ruling elite reflected the views of most of the Muslim inhabitants of the Empire. Indeed, the elements opposed to reform could always count on mobilizing the support of the populace by appealing to traditional values, "however vaguely formulated."[28] The reformers, who were also part of the elite, were not ready to enlist the support of the lower classes by trying to involve them in the process of rule.[29]

With the overthrow of Selim III, the triumph of the opponents of reform seemed complete. All the surviving partisans of the *Nizam-i Cedid* either fled or were hunted down and killed. But Selim's successor, Mutapha IV was to rule for only a little more than a year. An uprising against him led by provincial notables who were supporters of Selim, if not necessarily partisans of his reforms, resulted in the assassination of both of them. This brought to the throne the last surviving male member of the House of Osman, Mahmud II.

Mahmud had been a friend and confidant of his uncle, Selim III, and shared his views concerning the necessity of military reform. If it was not immediately evident what should be the content of these reforms or how he should go about effecting them, his uncle's fate indicated the need for prudence. According to the available evidence, Mahmud hoped at first to be able to reform the armed forces with the least possible disruption, purging the Janissaries and gradually restoring them to their historic discipline and military effectiveness. As early as 1809, two years after his accession, Mahmud endeavored to reestablish a measure of sultanal control over the corps. There was an immediate outbreak of violence, so he desisted. During the next decade he made two further attempts to do the same thing, also unsuccessful.[30]

The event which more than any other impelled Mahmud to undertake a drastic and concerted effort at military reform was the Greek revolt of 1823. As was usual in the event of war, most of the Janissaries refused to march, thereby obliging the government to raise levies of untrained and ill-disciplined troops. These were useless against the equally untrained but nevertheless ferocious Greek guerrillas. Only when troops dispatched by Muhammad Ali, the Ottoman governor of Egypt, entered the fray, did Ottoman fortunes improve. Muhammad Ali was then in the process of creating for himself a quasi-independent situation in Egypt on the basis of a well-disciplined, Westernized military force. In this army was incarnated Selim's vision of a *Nizam-i Cedid,* the name Muhammad Ali also took for his forces. Mahmud recognized the danger to the already tenuous unity of the Empire inherent in the possession of such a force by an ambitious, possibly dissident vassal, and he no doubt saw the necessity of creating something comparable.

Mahmud's initial approach to the problem was, like his earlier efforts, a moderate one. He planned to incorporate members of the various Janissary units into new, more efficient formations that were still organized according to essentially the same principles as the old corps.[31] Harboring no illusions about how difficult or complex this task might prove to be, Mahmud made very careful political preparation prior to launching his program, especially with respect to the religious authorities. Traditionally the ulema had tended to be the allies of all those who wished to see no changes, but in the eighteenth century some of them had come to appreciate the views held by the advocates of reform. Relations between the religious authorities and the Janissaries had also become strained in recent years, leading at one point to an armed clash in the streets of Istanbul between the latter and some theological students.[32] Mahmud was careful to favor the promotion within the ranks of the ulema of men known to be his partisans, while he was assiduous in the performance of his religious duties. Good relations with the ulema gave him a measure of control over public opinion, since preachers were a major channel of propaganda. Mahmud also took pains to see that his supporters were introduced into responsible positions within the Janissaries.[33]

At a series of meetings beginning in late May 1826, Mahmud brought together the leading civil, military, and religious personages of the Empire in order to work out the details of his proposed reforms. When queried as to whether the reforms were consonant with Islamic doctrine, the members of the ulema there present stated it as their unanimous opinion that Muslims had the duty to acquire military science. The projected military units were to be outfitted with new weapons and uniforms and trained according to a mode of drill, which, it was carefully pointed out at the moment of its introduction, was not European, but rather had been taken from the Egyptian forces of Muhammad Ali.[34] Even though there were already signs among the rank-and-file Janissaries of their discontent at the prospect of being obliged to adopt what they saw as the military styles and tactics of the infidels, preparations went quickly ahead.[35] Mahmud was also prepared for the eventuality that the members of the Janissary corps would reject his reforms. Of crucial importance was the availability of loyal troops, and here Mahmud relied on men from the bombardiers and the artillery. Both corps harbored some grudge or other towards the Janissaries. When on June 14, 1826, the traditional signal for a Janissary uprising was given, the overturning of soup kettles in the barracks, Mahmud had adequate forces at hand and the resolution to use them. He would not endure the fate of his uncle who had thought to avoid trouble by giving in to the conservatives.

At the moment of the rebellion, both the Janissaries and the troops loyal to the sultan tried to evoke the support of the local population. Mahmud was in a more advantageous situation in this regard, having already won to his camp a large segment of public opinion through his influence with the leadership of the ulema. At the same time the arrogant and overbearing behavior of the Janissaries in recent years had been such as to alienate the populace.[36] In the face of evident public antipathy towards their cause, the Janissaries could do little but withdraw to their barracks, all the while refusing to accept the plans of the sultan. When last-minute efforts to reason with them were greeted by howls of defiance, troops loyal to Mahmud brought cannon up to the fortified gates of the barracks square and opened fire on the masses packed inside. Over the next half-hour of carnage thousands of the rebels perished. Thousands more were hunted down throughout the Empire and either were killed or exiled.[37] Mahmud followed up his coup de force with an edict formally abolishing the Janissary corps.[38]

The sultan next moved against the Bektashi dervishes. The connection between the Janissaries and the Bektashis was of too long standing for the latter to escape unscathed in the general settling of accounts. With the backing of the ulema, no doubt happy to see so heterodox and widespread a movement weakened, Mahmud had the leading Bektashis either executed or imprisoned, while their meeting places were turned over to other religious orders when they were not destroyed.[39] Because the Bektashi order had so many adherents within the Empire, especially in Anatolia, Mahmud could not have proscribed it without arousing potentially unmanageable opposition. Nevertheless, its influence was for the moment drastically reduced.

By the same decree which abolished the Janissaries, provisions were made to replace them with a new force, the *Asakir-i Mansure-i Muhammadiye,* "the victorious soldiers of Muhammad". Later in the summer orders were promulgated organizing this new force along the lines of the *Nizam-i Cedid,* with twelve thousand men to be stationed in Istanbul and a comparable number in the provinces. Mahmud may have intended that the new *Mansure* forces be the nucleus around which the entire Ottoman army would be reformed, but for the moment they did little more than replace the discredited Janissaries.

The office of *Aga* or Commander of the Janissaries was eliminated and in its place was instituted that of *Serasker,* who was to have responsibility only for the new formations. Since the title did have a traditional prestige associated with it, this indicated that its incumbent was to be considered as holding a position distinctly above that of the other corps commanders. Over the next decade, the *Serasker* assumed an increasingly large share in the direction of military policy, taking on the attributes of a minister of war, although that may have been mainly a result of the forceful personalities of the first two men to hold the post.[40] Whatever his long-term plans for military reform, Mahmud did not care to act precipitously, for he had too many immediate problems to deal with, including the war with Russia which broke out in 1828 and the ongoing Greek insurrection, not to mention the problems posed by Muhammad Ali in Egypt. So it was not until 1831 that Mahmud revoked the remaining *timars* and thereby did away with the militarily useless traditional cavalry corps.[41] In so doing, he liquidated the institutional vestiges of Ottoman feudalism. Many of the other fighting corps would remain as independent organizations within the military establishment until 1838.[42]

Mahmud launched his program of military reform in order to save the Empire against its numerous adversaries, foreign and domestic. The defeat of the Ottoman forces first at the hands of the Russians and then against the troops of Muhammad Ali three years later in the first Syrian War demonstrated that more was required than simply the destruction of the Janissaries and the formal resurrection of something like the *Nizam-i Cedid* if the sultan's hopes were to be fulfilled. The hurriedly organized armies would have to be trained in modern military techniques as well as being provided with a cadre of officers capable of overseeing that training and then of leading the troops in battle. Such a cadre could not be produced in a hurry. The government sought officers wherever they could be found, even among the surviving remnants of Selim's *Nizam-i Cedid,* although after a lapse of some two decades they were more often than not superannuated.[43] In the end, Mahmud was forced to rely on men from the traditional Ottoman upper classes or from his personal slaves. None of these people had either the military skills to command European-style tactical formations or the inclination to acquire them.

The Ottoman government tried to hire European officers to assist in establishing the new forces, but for a number of reasons not much was accomplished here, at least for several years. Possibly unaware of the full gravity and complexity of the problem of military reform and also convinced of their own martial genius, the Ottomans were reluctant to look to Europe for anything but superficials.[44] In their turn, most of the European powers were not anxious to render the Ottomans the kind of military assistance they wanted. As long as the Greek revolt was still going on, most European states favored the insurgents, while France, the country with traditionally the greatest interest in the Near East, found that her concerns were shifting with the evident decline in Ottoman power.[45] France had become the sponsor of Muhammad Ali, and was disinclined to render assistance to his suzerian and chief adversary, the sultan. Thus only a trickle of foreign specialists took service under the Ottomans with the result that "Europe's active contribution to the training of the new army was very limited."[46]

In the end, the Ottomans were most comfortable utilizing officers from Prussia since it was the European power with the least overt interest in their affairs. Among the small cadre of Prussian officers assisting in training the Ottoman army was a young lieutenant, Helmuth von Moltke, who stayed from 1835 until 1839.[47] All the efforts undertaken by him and his fellows to improve the condition of the army would seem for the moment to have been fruitless, for in yet another confrontation with the forces of Muhammad Ali, the Battle of Nezib in June 1839, the Ottomans were again routed. Mahmud died just before he would have learned of this catastrophic culmination to his reign.[48]

The growing needs of the reorganized armed forces required that the resources of the realm be managed in a more systematic fashion. This in turn led to an increase in the power and authority of the sultanal government, thus ending two centuries of political and administrative decentralization. Indicative of the reestablishment by the central authorities of their prerogatives was the first Ottoman census and cadastral survey of modern times, carried out in 1831. Its immediate objectives were to facilitate conscription and taxation: "men for the new army and money to support it."[49]

Along with the growth in the power and the pretensions of the authorities at the center during the reign of Mahmud II, pressure developed for a more efficient organization of the machinery of government. The armed forces, as the first elements within the state to be set up along more modern or European lines, furnished a model and a testing ground for similar developments in other parts of the Ottoman system of administration.[50] Those departments most closely associated with the armed forces would seem to have been the earliest affected. The offices of the *Serasker* and the Grand Admiral were organized as functional ministries in 1833, while it was from the specialized bureau originally set up to oversee the financial requirements of the new forces that a European-style ministry of finance eventually evolved. The transformation of the army treasury into the chief financial organ of the Empire also signified the fact that the most pressing fiscal concern of the state was the support of the army. In the last years of Mahmud's reign he spent some 70 percent of his revenues on defense matters.[51] A further example of the systematizing aspirations of the government was the reorganization of the bureaucracy into three divisions: the scribal, the military, and the cultural-religious. A given hierarchical rank in each of the three bureaucracies had its equivalent in the other two. The parallel between this and the Table of Ranks established by Peter the Great is to be noted.

Even though the Ottoman Empire now possessed at least the structure of a centralized bureaucratic state, this did not mean that the government would now function according to its imperatives. The old ways and attitudes persisted. All members of the bureaucracy were now supposed to be remunerated by salary rather than through the time-honored system of fees, or *bakshish,* but if they were happy to receive a salary, they were also reluctant to forego the traditional modes of compensation.[52] The government also planned to have state revenues collected locally in a regular fashion by its paid agents, until it discovered that there were simply not enough men available possessing either the competence or a sufficiently developed sense of honesty and public spirit to carry out this function. So the Empire had to continue to rely on the old, wasteful system of tax farms, which down through the nineteenth century were to prove chronically unable to raise the funds necessary to support the growing apparatus of state or the policies of the government.

To meet the needs of the new army in armaments and uniforms, the government sponsored the founding of the first modern factories in the Empire, but here the beginnings under Mahmud II were modest. For one thing, the Turks had never been much concerned with manufacturing and commercial affairs, and it would take some time before they were able to develop the requisite skills. To have undertaken anything more ambitious at present would have necessitated the employment of foreign technicians on a larger scale than Mahmud was willing to contemplate.[53]

Practically from the moment he destroyed the Janissaries, opposition to Mahmud's goal of reorganizing sultanal power on stronger, more autocratic grounds began to emerge. The Janissaries may have alienated a major portion of the inhabitants of Istanbul, but they nevertheless seemed to be an integral part of the natural order of things, and their elimination in so brutal a fashion had a disorienting effect on the populace. The group that was possibly most disturbed by Mahmud's actions was the ulema. Many of the higher ulema had seen the destruction of the Janissaries as leading to a relative increase in their own power. They began to have second thoughts as they came to realize the sultan intended that their power be drastically reduced. An obvious intimation of his intentions came in 1826 when the state took over the collection and expenditure of *waqf,* funds for religious purposes which had traditionally been under the control of the *ulema.* As for the members of the lower ulema, they disliked the heretical implications of Western-style uniforms and military drill. Many of them were no doubt secret adherents of the Bektashi rites or at least sympathetic to them.[54] There were also many having no affiliation or connection whatever with the ulema who were simply antagonized by the higher level of taxes necessitated by the development of a modern military establishment. The opposition forces that began to take shape were a source of bother to the government, but they ultimately posed no real danger. The destruction of the Janissaries had robbed them of an armed nucleus around which they could organize themselves for effective action, while the new army, despite all its failures in foreign wars, "proved deadly effective as an internal police force."[55]

The reign of Mahmud II, lasting over three decades, was among the longest of any sultan since the days of Suleyman the Magnificent. It was also of crucial importance in determining the evolution of the Ottoman Empire over the course of the next century, for in this reign the government took a number of decisive, perhaps irrevocable, steps in the direction of "modernizing" the Ottoman state, even if that had not been Mahmud's intention when he first came to the throne. All he had wanted to do originally was to adopt certain military techniques from Europe, integrating them into the already existing armed forces and disrupting the established order as little as possible. It was in effect a policy of traditional reform. When the intransigence of the Janissaries made impossible this relatively moderate program, they were forcibly eliminated, and the government was then obliged to create, practically on the spur of the moment, new formations to replace what had been destroyed. The process was carried out by trial and error, inevitably resulting in much confusion.[56]

Once major military renovation was undertaken, Mahmud and his successors were to find that a compelling logic prevented it from being carried out in isolation. The reform of the army necessitated reform and renovation in much of the machinery of state, if only to gain control over the necessary fiscal resources. The grim irony of the situation is that the political power and military strength of Europe, which the Ottomans were already unable to match in the eighteenth century, continued to grow in an exponential manner, so that for all their determined efforts, "the reformers were forever pursuing a rapidly receding target."[57]

A number of scholars have noted a possible parallel between the achievements of Mahmud and Peter the Great. To improve the state of their armed forces through the introduction of European military techniques and modes of organization was a major goal of both rulers, and the needs of the army became an essential driving force behind the reform effort in the other sectors of state and society. Where the Europeanized army created by Peter within a few years proved its worth on the field of battle and then in the decades following his death became the institution embodying, for better or for worse, that which was progressive and modern in Russian society, the military reforms of Mahmud II had no comparable effect. That Peter was more successful in his effort than was Mahmud is no doubt explicable in part by the Russian tsar's having been a stronger personage, cast in more grandiose mold. In these terms, a comparison between their accomplishments is of no great significance. Where it may be enlightening is with reference to the social and political milieu in which each had to operate. As Bernard Lewis has pointed out, "Peter was already an autocrat; Mahmud had to make himself one."[58] And here Mahmud succeeded only imperfectly. In Russia, the social and political structure was relatively simple, and there existed no institutions possessing sufficient autonomous authority to resist the unleashed will of the autocrat. Ottoman society, on the other hand, was a complex, heterogeneous affair, a multitude of overlapping, self-administering entities, such as religious communities and brotherhoods, local guilds, and urban and rural parishes.[59] The sultan and his government may be conceived of as simply one, albeit the most powerful, in a cluster of semiautonomous, pluralistic centers of authority. Singly or in tandem, they were quite capable of restraining the ruler's freedom of action.

Any era of major political and social change generally has two consecutive, interrelated aspects. First comes the period when the antiquated institutions of the past are finally destroyed, often over a relatively short space of time. This is followed by a much longer period when, with great effort, new ones are put together—frequently with whatever is salvageable from the old. Peter managed to accomplish both the first and a great deal of the second part of the task during his reign. Mahmud was unable to do as much. To have eliminated the Janissaries, to have abolished the remnants of Ottoman feudalism, and to have restricted the autonomous power of the ulema were not inconsiderable achievements, but Mahmud's efforts to replace what he had destroyed with new, effective institutions were not successful, at least not in his lifetime.

Serious concerted attempts to create a more modern institutional infrastructure for the Ottoman Empire began during the reign of Mahmud's son and successor, Abdulmecid. This period of reform, extending down to the coup d'état of 1876 and into the early years of the reign of Abdulhamid II, is known as the *Tanzimat*, the "Reorganization." The *Tanzimat* is generally considered to have begun with the publication soon after the accession of Abdulmecid in 1839 of the *Hatti-i Serif* of Gulhane. This document, known in English as the "Rescript of the Rose Chamber," was in effect a statement of the good intentions of the government. In it were set forth such principles as the security of the life, honor, and property of each subject and the equality of persons of all religions. The rescript also called for the abolition of tax farming and regular, orderly recruitment into the armed forces.[60] The implementation of these principles in terms of the enactment of effective laws was to turn out to be a slow, laborious process, so much so that another imperial rescript published in 1856 carried a statement of intentions almost precisely like the one promulgated in 1839.

The authors of the rescript of 1839 desired to introduce within the Ottoman Empire as many as possible of the attributes of a modern state, and one of these was equality before the law. But it was a principle which ran counter to the profoundest sentiments of the Muslim subjects of the sultan, particularly with respect to service in the armed forces. Through the centuries the Ottoman Empire had provided a remarkable example of toleration and coexistence, with Muslims and non-Muslims living in close proximity to each other and performing complementary roles in society. There was in effect a rough division of labor between them. Mercantile and business functions were seen as being the province of the non-Muslims, while political, administrative, and above all, military tasks were performed by the Muslims. Since the earliest days of the Empire, the army had been the exclusive preserve of the Muslims, in particular the Turks, and a natural attribute of their situation as the ruling element.

To incorporate non-Muslims within the army would have greatly enlarged the pool of available manpower, something attractive to the authorities. Certainly the Muslim populace could have welcomed any proportionate reduction in the burden of military service, but not if they had to serve in the ranks on a footing of equality with the infidels. For a Muslim to be under their command was, of course, unthinkable. Legal and political equality might be admitted as a principle in the abstract, but under those conditions it was not acceptable. As for the non-Muslims, they may have wanted greater legal and political equality, but not at the price of having to go into the army. When in 1855, during the Crimean War, a decree was enacted incorporating non-Muslims into the armed forces, a large number of the inhabitants of Rumelia in the Balkans were angered enough to threaten to flee to the mountains and even to quit Ottoman territory.[61] In Lebanon, adherents of the Catholic Maronite rite went so far as to send an envoy to France, in order to procure the support of that country, the self-proclaimed protector of the religious rights of Christians in the Near East, against the efforts of the Ottoman government to conscript them.[62]

Many in the government looked upon the common obligation to do military service as a step towards the creation of a real sense of

Ottoman citizenship. But the members of the non-Muslim communities were reluctant to sacrifice the clearly defined, if limited, special rights and privileges attached to their status in the name of some nebulous principle of Ottoman patriotism.[63] Since the earliest days of the Empire, the non-Muslim population had enjoyed a protected status in return for paying a poll-tax. According to the 1855 decree they were now to be absolved from this obligation if they would bear arms. When the Christians showed themselves reluctant to be recruited into the army, a new tax, the *bedel-i askeri,* was imposed on them, payment of which exempted them from military service. The principle of equality, enshrined in the imperial rescripts, was preserved in that Muslims were also able to purchase exemptions, but they had to pay a much greater sum than the Christians. This compromise, "theoretical equality and practical discrimination," may have been about the best solution obtainable at the time.[64] The historic division of labor between Muslim and non-Muslim would continue to prevail, with the result that all efforts to improve the military posture of the Empire during the next half-century would touch only the former. Not until the aftermath of the Young Turk Revolution of 1908 would another serious effort be made to incorporate non-Muslims into the army.

During the first years of the *Tanzimat* the new Ottoman army, heretofore set up in an ad hoc fashion, was given a more permanent form. Six, later seven, armies were organized on a regional basis, each under a field marshal, who had broad jurisdiction over the military affairs in a particular territory in the provinces, and who was directly responsible to the *Serasker.* This represented considerable progress in the reassertion of effective control by the central government, in that the provincial governors were thus deprived of their authority in military matters.[65] According to regulations promulgated in 1843, the troops were to be permanently organized into corps, divisions, and brigades. Every major unit was assigned to a particular locale from which it would draw its recruits and where its reserves would reside. In its overall organization, the Ottoman army most resembled that of Prussia and, on paper at least, was superior to the majority of European armies, for, as of that era, the latter still lacked permanently instituted army corps and divisions, as well as a rational system for incorporating the reserves.[66] The size of the army was fixed at some 150,000 men on active service and 90,000 in the reserves, with the soldiers being chosen by lot through a regular system of conscription and serving for five years.[67] Because of the existence of the reserve, the

Ottomans were able to come close to doubling the size of their forces at the outbreak of the Crimean War and to throw into line almost as many soldiers as the Russians.[68]

In their efforts to create an effective military force, the Ottoman authorities had one precious asset, the sturdy peasants of Anatolia. Sober, robust, inured to pain and privation, they possessed a stoical courage such that numerous observers looked upon them as among the best soldiers in the world. All they needed was to be well trained and decently led. A comparison between their abysmal showing in the Syrian War of 1839 against the forces of Muhammad Ali and their steadiness in a number of the battles of the Crimean War a decade-and-a-half later is indicative of the progress they had made.

Despite the admirable qualities displayed in battle by the ordinary soldiers, and despite the efforts of the *Tanzimat* reformers, serious deficiencies in the Ottoman army still persisted. Most observers would agree with Marshal de Saint-Arnaud, commander-in-chief of the French forces at the start of the Crimean War, that although the Ottoman army had a high command and common soldiers, there was not much in between, "no officers and even fewer N.C.O.'s".[69] One European serving in the Ottoman army in the 1860s was of the opinion that most of the officers were indolent, incompetent, and corrupt, with the result that inadequate attention was paid to training. As for the pay and administrative services, they hardly functioned at all.[70]

It was early recognized that one prerequisite for a corps of competent officers was an adequate system either for training them prior to entry into service or while they were on the job. Schools for the teaching of mathematics to officers in the navy and in the technical branches of the army had led a fitful existence since the eighteenth century, but there were no schools for the training of officers in the other branches of the army, or for the dissemination of what might be called military science. Under Mahmud II, attempts to launch some kind of modern system of military education had been tentative and experimental. After an abortive effort to set up a palace school to prepare future officers, Mahmud had organized special teaching companies in each unit and assigned to them soldiers with particular aptitudes. Most of the outstanding officers of the coming generation would receive their first formal instruction in military matters in one of these special training companies.[71] In 1827 the government also inaugurated the sending of a number of promising young men to Paris, London, and Vienna to complete their military education, a policy which Muhammad Ali had already undertaken a few years before for the officers in his army. There was, however, much opposition to this practice on religious grounds from conservative elements. Finally, in 1831, Mahmud initiated plans for the establishment of a military academy.

Opening its doors in 1834, the Military Academy was to become the leading technical school in the Empire, but during its early years, it progressed very slowly. There were few qualified instructors, and a total lack of books.[72] Then too, because of the inadequate preparation of the students, the Military Academy had to provide them with the necessary scholastic grounding before it could begin to teach its own proper curriculum. Although an elaborate course of study was outlined in the official press, the subjects actually taught, at least during Mahmud's reign, were limited to reading, writing, arithmetic, Arabic, and military tactics.[73] The first class did not graduate until 1847.[74]

Even before he established the Military Academy, Mahmud had opened a military medical school to train surgeons and health officers for his new army. The Military Medical School, and the Military Academy, along with the Naval and Military Engineering Schools inherited from the eighteenth century and the Staff College founded in 1849, were the first advanced institutions of secular education in the Ottoman Empire, and until the establishment of Civil Service School in 1859, they were the only ones. Not until the last quarter of the century would there be any significant number of schools established with the specific purpose of preparing people for a civilian career, while the first true university in the Empire would not come into existence before 1900. For most of the nineteenth century, advanced or higher education functioned in large part as an adjunct of the needs of the military no matter what career its recipients ultimately pursued.

The responsible authorities of the *Tanzimat* era early recognized that students were unlikely to profit from the curriculum at the advanced technical schools without some kind of prior training. This was something in no way provided by the traditional, religiously oriented system of schooling. The reformers also knew that they would have to move cautiously "so as not to affront the ulema openly."[75] There were numerous committees concerned with the matter, but financial stringency impeded the rapid development of secondary education. Mahmud II did make a start in this area, and during the *Tanzimat* the number of secondary schools slowly increased. These *ruisdiye*, or adolescence schools, were meant for young men between the ages of ten and fifteen. By the time of the Crimean War, there were some sixty of them in the entire empire attended by about 3,350 students. Considering the fact that there were some 16,750 students enrolled in the *medreses,* or advanced theological schools, in Istanbul alone,[76] the accomplishments of the *Tanzimat* in the field of secular secondary education were hardly impressive. They were certainly not sufficient to meet the growing needs of the Ottoman state, let alone those of the army.

In response to the slowness of the civil government in founding an adequate system of education, the military authorities in 1855 initiated their own network of schools. Over the ensuing years there were established in the district of each of the armies several *ruisdiye* schools. In addition, each army district was to have one advanced secondary school, or *idadi,* the intended purpose of which was to prepare students specifically for the work they would undergo at the Military Academy. Thus much of the population was given the opportunity to partake of secondary education at the schools set up by the military long before the civilian network was established or extended to them. The army also took the initiative in developing a system of primary or elementary schooling.

The magnitude of the educational effort sponsored by the military in the last half of the nineteenth century may be gauged from the number of schools in existence and the students enrolled there. By 1897 the army had twenty-nine *ruisdiye* or secondary schools in operation with a student population of some 8,250. As for the advanced military schools, they graduated about 7,300 students between 1873 and 1897. That figure was considerably higher than the one for those graduating from the various advanced schools for the civil professions, even if an increasing number of these had been founded during the same period.[77] In short, the military authorities had established a broad and comprehensive system of education. In all of the schools run by the army there was a military atmosphere which grew more pronounced the further one progressed. Students at all levels had to wear uniforms. In the advanced schools the teaching faculty was made up almost exclusively of army officers, but even in the secondary and elementary schools they constituted a sizeable portion of it.[78]

The military schools acted as an element of fermentation and stimulation in Ottoman education. They had no real roots in the traditional system and could embark more easily on an innovative path than could the state-sponsored civilian schools, which were more likely to be hampered by interference from the religious authorities. Mathematics, gunnery, and other technical subjects were the basis of the original curriculum in the advanced military schools, but later there were added more general courses, such as military history, languages, and literature—often from a military point of view. To produce the textbooks required for their curricula, the military schools set up their own translation offices and printing presses. On these presses were produced the first translations of European books that were used in Turkish schools.[79]

In the process of translating European works into a vernacular familiar to the future officers, the military authorities were obliged to deal systematically with the Turkish language as a medium adequate for instruction. No one had heretofore had to confront the problems of teaching it, since education had been focused primarily on the memorization of Arabic texts. The first Turkish grammars were written by faculty members at the military schools,[80] while a former Inspector of Military Schools composed the first basic text concerning literary genres in Turkish on the basis of a course he had taught to future officers.[81] The military pioneered the simplification of the Turkish script, leading eventually to the adoption of the Latin alphabet, and in general exceeded all other groups in mass educational activities.[82] It was in schools run by the army that one could find for the first time, however crudely conceived and implemented, the idea of a broad and general secular education as a desirable background and complement to more specialized training.

The reforms of the *Tanzimat* era may be taken as a classic example of what has come to be called defensive modernization. Like other societies considered by Europeans to be "traditional," the Ottoman Empire was faced with the prospect of gradual territorial dismemberment or outright conquest if it did not set about adapting itself to the political and military power possessed by its most obvious adversaries. Ottoman statesmen could not be unaware that the survival of the Empire in the nineteenth century depended as much on the unwillingness of the European powers to face the implications of its collapse as on its own strength.

The imperatives of defensive modernization caused serious strains within both state and society. In particular, the fiscal system, hardly adequate to meet the needs of so decrepit a structure as the Ottoman Empire had become in the eighteenth century, was pushed to the point of collapse when the state sought to outfit itself with those most expensive appurtenances of modernity, European-style armed forces. Although the government attempted to increase its revenues through the introduction of more efficient and systematic methods of tax collection, it lacked the effective authority at the grass-roots level as well as a sufficiently large corps of trained, reliable bureaucrats. The treasury might budget what were on the face of it reasonable sums to meet the growing expenditures of state, but it was chronically unable to collect the anticipated revenues.[83] When in the past tax revenues had been inadequate, Ottoman statesmen had resorted to such remedies as debasing the currency, or confiscating the fortune of some rich man, preferably a Jew or a Christian. In the early years of the *Tanzimat,* however, the government tried to meet the continuing budgetary deficit by issuing paper money or a kind of bonds. During the Crimean War, when its expenditure sky-rocketed, the government issued yet more bonds, forcing merchants and bureaucrats to purchase them.

The founding of the Ottoman Bank in 1856, mostly with British capital, marked a new step.[84] Heretofore, almost all the money raised by the government had been from the inhabitants of the Empire, but now a means had been created to borrow abroad. Over the next two decades, the authorities would make repeated use of these facilities at ever more ruinous rates of interest. Having traditionally left commerce to men from other ethnic groups, the Turks had only a limited sense of economic affairs.[85] They did not recognize the dangers of the financial course of action they had so easily embarked upon. Finally, in 1875, the Ottoman government had to cut the interest payments in half, causing consternation in European banking circles and leading to the establishment of the Ottoman Public Debt, which was an agency under French and British supervision to manage specified domestic sources of revenue in the interests of the European bond holders. In other words, what had originally begun as an effort to defend the territorial integrity of the Empire through the creation of modern, European-style armed forces led to a concomitant growth in the power of the central authorities, along with the neutralization or destruction of any groups capable of restraining their potentially arbitrary, irresponsible behavior. Imminent financial catastrophe was the result, followed by the setting up of the Ottoman Public Debt, which in itself represented an abridgement of Ottoman sovereignty.

Concerning what measures should be taken to rescue the Empire, there was no consensus among those who were opposing the policies of the autocratically inclined reformers of the *Tanzimat.* The ideological vision of the most vocal and articulate of the opposition groups, the so-called Young Ottomans, was an imprecise amalgam of Muslim traditionalist values leavened by current European liberal ideals. European societies seemed able to experience political liberty at the same time as they wielded growing economic and military power. The means through which this was accomplished, or so it seemed, was a system of parliamentary government.[86] That something similar might be the salvation of the Empire was the hope of many of the young Ottomans.

A series of major mishaps and calamities in the mid-1870s allowed the opposition to see its political vision realized, at least in part and momentarily. In addition to the growing financial crisis, there were crop failures resulting in famine and widespread domestic violence. Then in 1875, rebellion broke out in Bosnia-Herzogovina and spread to Bulgaria. It provoked the Ottoman government to brutal repressive measures, thereby arousing antagonism everywhere in Europe. The reaction was especially intense among the Russians and led to their declaring war on the Empire in 1877. The Ottomans were defeated in this war, a defeat resulting in the loss of considerable territory in the Balkans. Still, the Ottoman army did perform in a very creditable way, notably during the siege of Plevna. In the face of the worsening situation, the government of the capricious sultan Abdulaziz, who had succeeded to the throne in 1861, seemed incapable of doing anything. On May 10, 1876, riots by theological students in Istanbul set in motion a train of events leading to the overthrow of Abdulaziz three weeks later. His successor, Murad V, proved to be mentally unstable and was deposed within three months, to be followed on the throne by Abdulhamid II.

Opposition groups like the Young Ottomans played little part in the events leading to the overthrow of Abdulaziz. Rather it came about as the result of a palace revolution engineered by Midhat Pasha, one of the leading figures in the civil administration, along with the two high-ranking army officers, the *Serasker* and the head of the Military Academy. They and other high officials were convinced that the irresponsible policies of Abdulaziz were ruining the Empire.[87] Midhat would have preferred the overthrow of the sultan to come about in as "constitutional" a manner as possible, the result of growing political pressure from below led by the ulema, although the military men favored a more direct and speedy approach. In the end, there was not enough time for a gradualist, "constitutional" solution, so it had to be a coup d'état, one in which the forces utilized came from the Military Academy.[88]

With the apparent approval of Abdulhamid, the leading figures in the deposition of the two previous sultans were next able to introduce a constitution. A liberal document on the model of the Belgian Constitution, it was promulgated at the end of December 1876. Over the succeeding fifteen months, parliament sat for two separate sessions and was then dismissed by the sultan. It was not to meet again for more than three decades. Whether the 1876 experiment in constitutionalism, founded on a secular concept of Ottoman citizenship irrespective of race or religion, would ever have worked is debatable. It was never given a chance, for the sultan was determined to preserve the fundamentally Islamic cast of the Empire. At the same time, Abdulhamid considered himself to be a reformer. Distrusting the honesty and competence of the average Ottoman politician and believing that a parliament consisting of such types would only delay the enactment of necessary legislation, he preferred an authoritarian approach to affairs of state as being more efficient. In keeping with that point of view, Abdulhamid established methods of personal control over the behavior of the civil servants, which in combination with the centralized system of administration developed under the *Tanzimat,* gave him greater real power in internal affairs than had been possessed by any sultan over the past three centuries. Thus armed, Abdulhamid was able to reorganize and consolidate what remained of the Empire after the unfortunate war with Russia. He also managed to bring to a conclusion most of the reforms initiated in the previous decades. In that sense, Abdulhamid may be considered as "the last man of the *Tanzimat*".[89]

Abdulhamid was one with most of the mid-nineteenth-century reformers in his belief that the Empire could be governed best as an autocracy. Autocratic government presupposed that there be a sizeable corps of trained, competent, and, it was to be hoped, obedient servitors to assure the regular functioning of the apparatus of state. One major purpose of the schools established under the *Tanzimat* was to provide those servitors, men who were in effect a new caste within the Ottoman ruling elite. The founders of the Ottoman system of secular education saw it as a vital element in the reform, and thus the preservation, of the Empire. Implicit in their view was the continued existence of the present structure of social and political power, albeit with a few necessary adjustments, but the very process of educating the new administrative caste was to create problems for the governance of the Empire under Abdulhamid.

The ruling authorities found it difficult to limit the knowledge acquired by those enrolled in the advanced schools to purely technical or professional matters. In the course of their schooling the students usually learned at least one European language, while a number of them went to Europe for varying periods of time to continue their education. In 1865 an Ottoman school was established in Paris to prepare Turkish students for entrance into St.-Cyr, Polytechnique, and the other Grandes Ecoles. By being introduced to the European academic milieu, the students were swept up in the "self-sustaining effervescence" of the European world of ideas.[90] They could also thereby develop a new frame of reference in accordance with which they might seek, if they were so inclined, to elucidate and understand current Ottoman realities. By their schooling, then, not a few of the students found themselves increasingly critical of the ruling establishment and even disaffected from it. They were examples of a phenomenon to be seen again and again not only in the Ottoman Empire but in many societies subjected to the powerful influence of European ways and institutions.

In a traditional society the training reserved for the elite, whether membership in it be determined by birth or by indigenous concepts of merit, is generally consonant with the fundamental norms of that society. Even if those in the elite receive a disproportionately large share of the material advantages to be obtained in their society, that favored situation is perceived by the great mass of the nonelite as being legitimate. Elite and nonelite perform complementary roles; both inhabit the same moral universe.

That is not the case in societies like the Ottoman Empire, where some in the ruling establishment have introduced administrative methods and governmental organizations modeled upon European practice. Those trained to operate the new modern, or European-style, machinery of state may envision the process of government, even the workings of society, in ways contrary to the assumptions of the great mass of the populace concerning these matters. To that degree, the men in the new elite feel estranged from those they are meant to rule. And because the populace has little appreciation for the modern modes of government or administration, not really accepting the criteria by which the new elite has been designated, the latter has little legitimacy in its eyes. The directives of the new elite are followed primarily because it controls the means of coercion. And for their part, the members of the new, technically trained administrative caste produced by the schools are not necessarily comfortable with their recently gained status, despite the perquisites attached to it, since that status is not really legitimate in the eyes of the masses from which they spring. In the Ottoman Empire the most important group subject to these ambivalent feelings was to be found in the junior ranks of the officer corps.

By the last decades of the nineteenth century, the younger, school-educated officers came increasingly from the provincial lower middle classes. The military schools and the army itself formed a vehicle par excellence of social mobility, bringing to positions of responsibility within the Empire those from outside the traditional ruling element. In addition to being estranged intellectually and psychologically from their natal roots, they also had little in common with most of their fellow officers, in the main promoted rankers.[91] Their social identity was founded above all on their education and on their consciousness of being members in a new professional caste, one which was evolving its own corporate sense of purpose and its own norms of conduct. The new caste of school-trained officers was not really integrated into the fabric of the Ottoman governing establishment, despite its growing importance for the Empire. Unlike those from the traditional elites, these officers had less of a stake, psychological or social, in the status quo. They were thus willing to contemplate changes in it, if such seemed necessary according to criteria dictated by their professional outlook.

The state of mind of the armed forces was a matter of considerable importance to Abdulhamid. As a ruler with autocratic predilections, he must have felt misgivings in the presence of any more or less autonomous center of power and authority within society. Abdulhamid could also not be insensible to the role played by a few high-ranking officers in the overthrow of his predecessors and in his own elevation to the sultanate. The better to exercise his control over the army, he moved to reassert the traditional prerogatives of command held by sultans in the past. The authority of the *Serasker* vis-à-vis the army was reduced, while Abdulhamid wielded command directly through a series of permanent commissions. He also devoted much effort to making sure of the army's loyalty to himself, personally approving the promotion and assignment of every officer. Palace spies were present in all military units, while both officers and men were encouraged "to report on their fellow soldiers and upon their superiors."[92] With a system of promotion which came to work primarily through favoritism, any person at all concerned for his career attached himself to the entourage of some influential officer. Those stationed closest to Istanbul had the best chance to be promoted; assignment to a distant provincial garrison was tantamount to remaining permanently in grade. The more ardent, capable officers found themselves thwarted in the performance of their professional duties by the suspicions of the sultan. If a young officer had not become disillusioned during his time at school, he was likely to be pushed in that direction by what he saw of the realities of military service.

For all his suspicions with regard to the officer corps, Abdulhamid had a sufficient sense of political reality to appreciate how much the well-being of the Empire depended upon the effectiveness of the army. A number of measures were taken by the Hamidian regime to improve its situation with regard to weaponry and fortifications as well as recruitment. Abdulhamid was also willing to avail himself more openly of the services of military advisers from abroad than his nineteenth-century predecessors had been, even though he constantly sought to emphasize the Islamic character of the army. In 1886, Colmar von der Goltz, a German officer who possessed a considerable reputation as a military thinker, arrived in Istanbul, there to spend more than a decade as adviser to the sultan.[93]

Throughout the Hamidian era, the armed forces continued to consume a very large share of the state revenues, well over half of what was left after the annual interest and amortization on the debt had been paid.[94] Since the sums collected through taxation continued to be insufficient, the government was obliged to borrow and to resort to other expedients. The pay of the officers was almost always several months in arrears; as for the troops, they were seldom paid at all.[95] Needless to say, these practices worked to counteract the military improvements introduced by the sultan.

Discontent among students and opposition to the policies of the government existed from the beginnings of the Hamidian regime. They grew more acute in the late 1880s, when the sultan attempted to cope with the persistent budgetary deficit by reducing the staffs in both the government offices and in the army. Reductions of this kind directly affected members of the educated classes, who now found themselves without jobs, and who as a result were able to appreciate that much more vividly the iniquities of the Hamidian system.[96] A particularly serious opposition movement began in 1889 among the students at the Military Medical School. It grew rapidly and spread to the students in other government schools.[97] The student conspirators also made contact with various groups of exiles living in Europe. An attempt by people in this movement to overthrow the sultan in August 1896 failed after the authorities had learned of the plans through an informer. Most of the conspirators were apprehended and sent off to prison or forced to reside in some remote corner of the Empire.

For the moment destroyed within the Empire, the opposition movement slowly reconstituted itself abroad. The existence of small groups of dissidents and intellectuals, generally living in comfortable exile, writing, agitating, and quarreling over both ideology and tactics, was not a new phenomenon in the nineteenth-century history of the Ottoman Empire. None of the most recent exile groups, who came collectively to be known as the Young Turks, can be seen as posing any real threat to the regime of Abdulhamid, and not a few of them, after a suitable delay, made their peace with the sultan and returned to take service under the government. Abdulhamid preferred to see them at home busily employed in a government office rather than agitating and conspiring in Paris or Geneva, and he generally made no trouble over their past opposition to him. When the opposition movement reawakened in the opening years of the new century, its chief focus would not be either among the students or the exiles in Europe, but rather among the junior officers of various provincial garrisons.

Throughout most of the years of Abdulhamid's reign, a primary mission of the armed forces was that of fighting rebels and trying to repress terrorism in Macedonia and Armenia. Some fifty percent of the active strength of the army was stationed in Macedonia following the most recent war with Russia.[98] What was frustrating to many officers was the realization that the government was either unable or unwilling to provide them with sufficient means to carry out their mission.[99] Money for material and training was short, while, as noted above, pay was perpetually in arrears. Even the least perspicacious officer could appreciate that it was questionable whether the army would be able to defend the Empire against its external or its internal foes.

When and in which garrison there began the conspiratorial activities leading ultimately to the Young Turk Revolution of 1908 cannot be ascertained with precision. In 1906 a young general staff captain named Mustapha Kemal, the future Ataturk, stationed with the Fifth Army Corps at Damascus, was one of the participants in the organization of the Freedom and Fatherland Society. He seems to have been instrumental in establishing a branch of this society in Salonika, the city of his birth and one of the major garrisons of the Third Army Corps, when he visited there later the same year, but he was simply one of many, and by no means the most significant.[100] The military conspirators also made contact with some of the groups in exile, in particular the Committee of Union and Progress, C.U.P., the name eventually assumed by the leading military organization. Although the Salonika officers maintained a connection with the Union and Progress group in Paris, it was primarily as a matter of convenience. In the conduct of their affairs, they maintained a very independent stance.[101]

A number of factors favored the rapid growth of conspiratorial activities in Macedonia, especially Salonika. It was the most cosmopolitan city in the Empire, open to European intellectual currents.

The presence there of an international force of gendarmerie, set up at the behest of the European powers following a long series of nationalistically motivated incidents of terrorism, and intervening in what should have been a purely domestic concern of the Ottoman government, namely the maintenance of order, was in itself a standing reproach to the increasingly incompetent tyranny exercised by Abdulhamid. It was also a source of humiliation to the Ottoman officers, already exasperated by the loss of so many comrades in the interminable struggle against the terrorists. In that fertile ground the spread of the branches of the revolutionary society was so rapid that within a short time, "it was difficult to find a Turkish officer in all of European Turkey who was not pledged to overthrow the government he served."[102] With his elaborate network of spies and informers, Abdulhamid was aware that the Macedonian army was riddled with disloyalty and was, in fact, in a near treasonous state, but he could not take drastic measures to suppress the seditious activities without, in effect, "wrecking his army altogether."[103]

Beginning in 1906 and increasing through 1907, a number of mutinies took place in various parts of the Empire, triggered mainly by the wretched conditions in the army and the arrears in pay. Even though most of the demands of the insurgents were met, the tardy actions of the government did little to win back the allegiance of the junior officers. A large percentage of them were by now convinced that only some kind of change of regime could lead to an improvement in the military posture of the Ottoman state. In 1908 there were further mutinous outbreaks in Macedonia. Efforts on the part of the government to get control of the situation led, among other things, to the assassination of the general sent to take charge of the suppression of the dissidents. Concurrently, several officers decamped to the hills with some of their troops in outright resistance to the government. In late June and early July mutiny spread from the Third Army Corps to the Second Army Corps headquartered at Edirne and located much nearer to Istanbul. By the third week of July the army of European Turkey was effectively in revolt against the government. On July 23, the constitution, in abeyance since 1878, was proclaimed in various towns of Rumelia, while ultimata were issued announcing the intention of the army to march on Istanbul and depose the sultan should he not accept the restoration of the constitution. Under the circumstances, Abdulhamid had no choice but to acquiesce. Such then was the Young Turk Revolution of 1908. In its immediate origins it had little about it to distinguish it from the mutinies which had become so frequent in the recent past.[104]

The constitutional revolution of 1908 was not exclusively a military affair. Indeed the role of the civilian intellectuals in articulating the program of opposition to Abdulhamid has led one historian to characterize the 1908 movement as "primarily a political operation with only marginal military overtones."[105] Still, the movement of the intellectuals was able to accomplish little "until its leadership was taken over by the only efficient social force on the scene, which was to play a major political role in almost every Muslim country in the twentieth century—the nationalized, westernized, and secularist officer corps."[106] However much the civilian intellectuals, either at home or in exile abroad, did to provide ideological and political justification for the revolution, the main precipitating impetus behind it was the discontent of the soldiers at the military incompetence of the existing regime.

Overt intervention by the military in politics was nothing new in Ottoman history, but it would be misleading to draw too close an analogy between the actions of a few high-ranking officers in 1876 or the Young Turks in 1908 and those of the Janissaries in centuries past. When the Janissaries rebelled against the government, it was usually in defense of their corporate privileges, and if there was any ideological justification proffered, it was generally an advocacy of certain traditional, quasi-religious values and modes. The Young Turks, on the other hand, rose in the name of essentially new, secular concepts, "freedom and fatherland, the constitution and the nation."[107] These were not slogans likely to arouse any very profound emotions among the great mass of the people. The Young Turk rising was an elitist affair, with no real broad popular support, even though the protagonists sprang from the people and not from the traditional ruling strata. The officers involved had to do a great deal of propaganda work among the men in their units to convince them that they were taking arms not against the sultan, still very much a revered figure and possessed of a religious aura, but rather against his corrupt advisers. This may have helped to assure the support of the rank and file of the army once the revolution broke out, but it also precluded the dethronement of Abdulhamid after it was successful.[108]

In addition to a desire to recover the military efficacy of the Empire, the restoration of the constitution was the one thing upon which all those immediately involved in the revolution could agree. This political goal was achieved with remarkable ease and little bloodshed. That there were serious differences among the revolutionaries over what should be the future shape of the Ottoman polity was for the moment hidden, but from the start two main tendencies existed, one liberal and the other nationalist. The military were represented in both camps, although they were more likely to be found with the latter. The liberals had a vision of an Ottoman fatherland with a common citizenship in which all ethnic and religious differences would be reconciled through generalized participation in the political process. It was a vision with roots in the two reforming imperial rescripts of 1839 and 1856. The structure of the Empire according to this vision would be rather decentralized and federal. Opposed to the liberal ideal was that of the nationalists. They favored an empire with a more centralized structure, and if they had some vision of an Ottoman citizenship, they were also reluctant to see the Turkish elements lose their historically dominant role. The army officers were preponderantly Turkish in nationality and had an understandable predisposition towards, and indeed a vested interest in, a strong centralized state.

Following the election of a new parliament in December 1908, the Empire was governed as a constitutional regime for some four years. During that period, the military, having provided the major force in effecting a restoration of the constitution, withdrew from active participation in politics, but they continued to exert an influence from behind the scenes. Thus the budget for the armed forces rose at an even faster rate than did the budgets of the other branches of the state, and special efforts were made to repair some of the material neglect in the armed forces which had occurred in the last years of Abdulhamid.[109] In general, the new constitutional regime made a modest but respectable start, but one insufficient to meet the expectations of various groups within the state. For the restored constitutional government to have begun to function as hoped, there would have had to be a period of relative calm and an absence of undue pressure on the new institutions. Those preconditions did not exist.

In addition to periodic domestic crises there was an almost unbroken series of disasters abroad. In 1908 Austria-Hungary ended its thirty-year occupation of Bosnia-Herzogovina, still nominally an Ottoman possession, by annexing the territory. Next, Bulgaria declared its full independence, thus terminating the last shreds of Ottoman suzerainty there, and Crete, also nominally Ottoman, announced its union with Greece. Then in 1911 Italy declared war over Tripolitania, while 1912 saw the outbreak of the Balkan Wars, leading to the loss of almost all that was left of Ottoman holdings in Europe. One factor probably contributing to the Balkan Wars was the fear on the part of different ethnic and religious groups within the Empire, and their supporters beyond its borders, that the new government was about to initiate a more forceful policy of centralization, to the benefit of the dominant Turkish elements. Initiated to assure the integrity of the Ottoman Empire, the Young Turk Revolution may thus have precipitated the train of events leading to a major reduction in its territory.

The assault of the Balkan states on the remaining Ottoman lands in Europe discredited efforts by the liberals to build the Empire on the theoretically stronger grounds of voluntaristic cooperation among the different ethnic groups. It is to be doubted whether anyone in the governing elite ever really considered giving to the non-Turkish or Christian elements a political and military role commensurate with their numerical importance. Nevertheless, in accordance with the spirit of equality embodied in the constitution, a law was enacted calling for the conscription of Christians into the army on the same basis as Muslims, along with the gradual abolition of the exemption tax they had heretofore paid. If the liberals hoped that such military service would help to build a sense of secular commitment to the Ottoman Empire on the part of non-Muslims, the more nationalistically inclined, especially among the officers, saw it primarily as a means to increase the military manpower of a state under assault from all sides.

Any realm is more likely to be able to maintain itself against external foes if it possesses a measure of unity or homogeneity, be it ethnic, social, or cultural. At the very least, an army can contribute to that homogeneity by suppressing any overtly dissident movements. This the Ottoman army did on numerous occasions both in the nineteenth century and before. The military forces may also contribute to building homogeneity within a society by acting as an agency of socialization through which recruits can be indoctrinated in a single uniform way and brought to accept the goals and norms of the society as they are articulated by the elite. This was one of the functions of universal military service in late nineteenth-century Europe, but a number of factors prevented the Ottoman armed forces from ever performing an analogous role. Where in Europe the concept of the "nation-in-arms" was founded on an already existent loyalty to the state, in the Ottoman Empire there was no state in the modern, European sense, "but rather groupings within the Empire based upon religion, race, or culture."[110] The structure of political and social power reflected the realities of the Ottoman conquest over four centuries, and the ruling elite was ultimately unable or unwilling to change it in such a way as to encourage in the non-Turkish elements a sense of involvement or commitment. The Young Turk officers who instigated the revolution of 1908 thought of themselves as Ottomans and had a vision of an Ottoman national state, but it was one with a predominantly Turkish bent to it and thus not likely to elicit much enthusiasm from the other ethnic groups. Under the circumstances, any attempt to institute universal military service would have contributed little to increasing the strength of the Empire above and beyond providing more recruits for the armed forces.

If the military had sought to avoid overt participation in the political process, following the overthrow of the Hamidian regime, the troubles which beset the Empire more or less continually from 1908 on brought them back into the arena. Thus a group of liberal-leaning officers intervened in 1912 against the political domination of the nationalistic C.U.P.[111] Some six months later, as the First Balkan War was coming to its catastrophic conclusion, a small band of soldiers, adherents of the C.U.P., staged a countercoup, bringing to an end the last vestiges of constitutional legality and ushering in what became essentially a dictatorship under the triumvirate of Enver, Talat, and Cemal. They would rule over the Empire for the next five years.

The Ottoman Empire entered World War I on the side of the Central Powers, prodded, it would seem, by the Germanophile tendencies of Enver. Although the performance of the Ottoman army in the war gave on more than one occasion undeniable evidence of the progress achieved in military matters down through the nineteenth century and particularly in recent years, the Empire was in the end unable to support the pressures of modern war, one demanding both a high degree of commitment on the part of the populace and an effective machinery of state.

Defeat in World War I provided a final, brutal solution to many of the problems of ethnic and religious integration which had bedeviled Ottoman statesmen for the past 100 years. What remained of the lands acquired by force of arms over the centuries was sheared away, leaving behind little but the basic, ethnically homogeneous nucleus of Anatolia. In the midst of defeat and decomposition, the army came to play a vital and creative political role. Of all the institutions of the Ottoman Empire, the army had benefited more than most from the effects of Europeanizing reform. If such a program of reform had been unable to save the Empire from ultimate collapse, it did at least endow a large number of the officer corps with a "modern," secular outlook. Possessed of resilience and public spirit, they furnished the vitalizing impetus in the creation of a new political existence for the Turkish people.

The founder and animator of the new regime was probably the most capable general in the recent war. Mustapha Kemal was to prove to be one of the truly remarkable statesmen produced by any nation in the twentieth century, and among the qualities he brought to his self-appointed task of national reconstruction and regeneration were soldierly ones—discipline, realism, and a sense of order. It is noteworthy that Kemal and his associates did not found a military regime, but rather an avowedly civilian one. If their dream was not to be fulfilled in quite the way they envisioned it, still they came closer than most who have tried to reshape a traditional, non-European society in accordance with Western norms.

4 *The Egyptian Army of Muhammad Ali and His Successors*

A prominent characteristic of the Ottoman Empire during its centuries of decline was the progressive loss of control by the central government over many of the outlying provinces. Always a danger in any system of patrimonial administration, the process had gone so far by the end of the eighteenth century that certain areas, including Albania, Mesopotamia, and much of North Africa were now quasi-independent satrapies. A major motive underlying the program of military reform launched by Selim III in the 1790s had been not only to defend the Empire against its external foes but also to reassert the power of the central authorities in the face of these particularist tendencies.[1]

If, during the course of the nineteenth century, the Ottoman government was able to regain control over many of the provinces in question, it had only limited success with regard to the North African lands. Algiers was lost through conquest by the French in 1830, while Tunisia maintained its independence from Istanbul until 1881, at which point France declared it to be a protectorate. As for Egypt, that country was able under the leadership of its governor, Muhammad Ali, to fend off the reimposition of Ottoman suzerainty so effectively that by 1840 it had become to all intents and purposes an independent kingdom ruled by its own dynasty and possessing indigenous institutions of state, the most noteworthy of which were its Europeanized military forces.

In Egypt more than any other extra-European society of the nineteenth century, the demands of a modernized army provided the driving force behind the emergence of a new social and political system. As one noted Marxist scholar has declared:

> For Muhammad Ali the army was neither an instrument of power, . . . a prime element of the state, nor a department of government: it was everything, the pivot of national life. . . . With the army as his starting point, Muhammad Ali constructed a state and restored life and strength to millenia-old Egypt.[2]

In addition to having a profound effect on the evolution of Egyptian society, the army also provided a spur to military reform in the Ottoman Empire itself. Without the example provided by Muhammad Ali as to the efficacy of Europeanizing military reform, it is doubtful that the sultan would have been able to overcome the conservative forces of resistance within Ottoman society as rapidly as he did, if at all.

Egypt had been a part of the Ottoman Empire since 1517, when an army led by Selim the Grim had defeated the Mamluk sultanate, ending its 250-year period of independent rule. The Mamluks, a typical Islamic force of slave soldiers, were incomparable cavalrymen. Military individualists that they were, they could not cope with the disciplined tactics of the Janissaries in their golden age.[3] But, even if defeated, the Mamluks as a sociopolitical phenomenon did not disappear from the scene. Egypt was far from the main center of Ottoman power and was never incorporated into the ordinary system of administration. Rather an Ottoman army of occupation was permanently stationed there. It eventually lost its distinct identity, along with its discipline, and became integrated into Egyptian society through intermarriage and through its members taking up various trades. The officers of the occupying forces imported slaves, mainly Circassians from the eastern shores of the Black Sea, who were then trained in the military arts. Thus, a caste of Mamluks came into existence once again. As the military capabilities of the Ottoman forces declined, the defense of Egypt increasingly became the responsibility of these new Mamluks. By the middle of the eighteenth century they had come to be the real rulers of the country.

Among the Mamluks the struggle for positions of power and domination was a perpetual one. The leader of a momentarily victorious faction would assume the title of *Shaikh al-Balad,* de facto ruler of Egypt.[4] If a *wali,* or governor, of the province sent out by the sultan happened to displease the *Shaikh al-Balad,* the sultan generally found it expedient to appoint a new one.[5] Between the Mamluk rulers of Egypt, a constantly renewed oligarchy of foreigners, and the great mass of people there was a tremendous gulf.[6] Under the Mamluks things went on as they had since ancient times, with the native-born Egyptians living under the domination of a foreign ruling caste.

The 1798 French invasion of the country under Napoleon Bonaparte marked an important turning point in Egyptian history. The Napoleonic episode was significant less for the permanent institutional and intellectual innovations brought about by the French than for the fact that the ascendency of the Mamluks was ended through their defeat at the Battle of the Pyramids.[7] In the confused situation created by Bonaparte's invasion and subsequent withdrawal, a number of factions struggled for control. Where political competition had in the past been limited to the Mamluks, it now involved not only the Mamluks but detachments of the British, remnants of the French, and representatives of the sultan. Ottoman authority in Egypt had all but disappeared by the end of the eighteenth century, but with the defeat of the Mamluks, followed by the withdrawal of the French expeditionary force, Selim III made an attempt to reestablish it. In 1801 he sent to Egypt a body of Ottoman troops, one significant contingent of which was made up of Albanian irregulars. Second in command of that contingent was Muhammad Ali.

Apparently Muhammad Ali owed his appointment to his having been a protégé of the local governor, although his family did have something of a military background. His father had been a commander of Albanian irregulars in addition to being a tobacco merchant, a career Muhammad Ali also followed.[8] By 1803 he had risen to the command of the Albanian troops because of his exemplary conduct in battle and because of the untimely death of his immediate superior. All the while, Muhammad Ali was maneuvering skillfully among the different competing factions in the political confusion characteristic of Egypt at that time, endeavoring to promote his own cause. Concerning his military ability there can be room for argument, but his political skills are beyond question. Muhammad Ali first allied himself with one element of the Mamluks against the Ottoman governor and then, having defeated the latter, with one party of the Mamluks against the other. At the same time, he was carefully building support with the religious authorities, or *ulema,* with certain of the indigenous mercantile elite, and with the populace of Cairo. He acted as the champion of the latter against the exactions of both the Ottoman government and the Mamluks. By 1805 his tortuous maneuverings had so well succeeded that in response to the appeals of the Cairo ulema, Muhammad Ali was named governor of Egypt by the sultan. By this time the power of the Ottoman authorities over the provinces had diminished to the point that all they could do was to legitimate the *fait accompli* when an official on the spot succeeded "in developing enough local power to impose some order on the surrounding chaos."[9] For a person without significant standing within the regular Ottoman establishment, who had arrived in Egypt only four years before as a subordinate commander in one of the contingents of the Ottoman expeditionary force, it was the fitting capstone to a remarkable ascent. He was to remain in power for the next forty years.

Muhammad Ali is generally reputed to be the person who decisively pushed Egypt, a relatively backward land, on the path to modernity. For this he is one of the more notable figures in the history of the nineteenth-century Muslim world, but at the start he may have been little more than an example of the typical military adventurer who on occasion was able to seize power and to achieve a large measure of autonomy in a peripheral area of the Ottoman polity because of the decay in the centralized institutions of state. Whatever the reasons for or circumstances of his elevation, Muhammad Ali soon came to look upon the lands over which he had been delegated authority in a highly proprietary fashion.

During the first decade of his tenure as governor, Muhammad Ali's situation was quite insecure, and most of his energies were devoted to consolidating it. His having received the official imprimatur of the Ottoman government was hardly sufficient to cow his former rivals and political adversaries in Egypt, for Istanbul was far away. Muhammad Ali also possessed no power base of his own beyond that provided momentarily by the adherence of the populace of Cairo and by the presence of the turbulent Albanian irregulars. Over the latter, his ascendency was analogous to that of a tribal chieftain rather than a military commander.[10] Although Muhammad Ali's primary instrument of power would eventually be his European-style army, he was able to begin to organize it only after he had been governor for some ten years, and the first fully trained units would not be ready much before 1823.

The most pressing concern immediately facing Muhammad Ali after he had been made governor was the paucity of his financial resources. He could not alleviate his shortage of funds by taking over any centralized organs for collecting tax revenues because those of the Mamluks had been either non-existent or of the most rudimentary kind. So in order to ameliorate his financial situation, he early resorted to a number of arbitrary acts, most notably the confiscation in 1809 of *waqf*, property the income from which had heretofore been tax-exempt since it went in theory to support the ulema. In so doing Muhammad Ali ran the risk of arousing the opposition of the religious authorities. With the cessation of Mamluk rule during the French occupation of Egypt, the ulema had by default enjoyed a correspondingly greater political influence among the people. As noted above, their backing had much contributed to Muhammad Ali's elevation to the post of governor in 1805. Acutely aware of their ability to create trouble, he sought to buy them off after his appropriation of *waqf* by assuming responsibility for the upkeep of the mosques and for the support of Al Azhar, possibly the most prestigious institution of learning in the Muslim world and the focus of ulema influence. At the same time he also moved to neutralize the potential political power of the ulema by discrediting and ultimately removing from office the chief shaikh at Al Azhar. On this occasion his tactics were such that the important figures at Al Azhar ended by being thoroughly divided among themselves and thus incapable of offering any real resistance to further encroachments by the secular authorities. It might be noted that the financial support Muhammad Ali proffered to the religious institutions turned out to be quite inadequate.

The Mamluks represented opposition of a different, more overtly dangerous kind. Discredited as they may have been by their defeat at the hands of Napoleon, they were still capable of acting as a serious military force, one that could hinder Muhammad Ali in the exercise of his power. Only after a political and military struggle lasting several years, in which he resorted to his usual tactics of allying first with one faction of the enemy and then with the other, was he able to subdue them, at least provisionally. Recognizing the ongoing potential of the Mamluks for sedition and rebellion, he decided to eliminate them once and for all as a threat. In 1811, on the occasion of his son Tussan being invested with the command of an expeditionary force which Muhammad Ali had been charged by the sultan to send to Arabia, he invited a few score of the leading Mamluks to a great ceremony at the Cairo citadel. After the ceremony, as the Mamluks were leaving the citadel via a narrow alley way, they were attacked by a detachment of Albanian troops and systematically massacred. With their leading elements thus eliminated, the remaining Mamluks were disorganized and powerless. Many of them fled to Upper Egypt where they continued to be an occasional nuisance. Others of the Mamluks, recognizing the futility of resistance to the new governor, took service with him.

Having rendered innocuous those forces most capable of opposing him, Muhammad Ali then set about gaining mastery of the potential resource of the country, represented above all by its cultivable land. Following the sixteenth-century Ottoman conquest of Egypt, that land had been assigned to the control of its new rulers as tax farms. In the course of the succeeding centuries, although still in theory leased from the state, the tax farms had come to be treated as hereditary and in effect private property by the lessors, chiefly Mamluks.[11] Thus economic power reinforced their political and military power.

In seeking to gain control of the arable soil in Egypt, Muhammad Ali might claim that he was doing no more than reasserting the traditionally understood proprietary or dominial rights of an Islamic ruler with regard to the land of a country. His pretensions here were somewhat questionable, for strictly speaking, he was no more than governor of Egypt, holding that office at the pleasure of the ultimate ruler, the Ottoman sultan. However tenuous the legal grounds for his actions, he proceeded to have a major new cadastral survey carried out to ascertain the extent of the country's agricultural resources. He then decreed the abolition of all the tax farms in Upper Egypt in 1812 and those in Lower Egypt two years later. Some of the former proprietors were to receive pensions from the treasury by way of compensation, but the sums involved were in no way commensurate with the income they had managed to derive from their tax farms.[12]

Following the abolition of the tax farms, the land was graded according to three categories of fertility and then divided among the cultivators, with every peasant receiving a holding consisting of a strip from each of the three categories.[13] About half the land in the country Muhammad Ali retained for himself and his family, but he intended to exercise control over how the remainder was utilized. Where peasants in the past, whatever the nature of their land tenure, had been able to cultivate their individual holdings as they pleased, obliged only to pay a certain amount to the local tax farmer, they now had to follow the directives of the government. The authorities ordered the cultivation of cash crops such as indigo, sugar, rice, flax, and above all, cotton, establishing monopolies to market them. The crops would be purchased from the peasants at a price set by the government, in general the lowest possible, and then sold abroad at a handsome profit.[14] In short, Muhammad Ali was running Egypt as a single, vast, capitalistic enterprise, on the profits from which he was able to assure the stability of the new regime and also to finance his growing imperialist ambitions.

Once he was in control of most of the land in Egypt, Muhammad Ali was able to use it for political as well as economic ends. Land provided a convenient means to reward those who had been faithful servitors and also those whose loyalty he hoped to win. Muhammad Ali was thereby laying the basis for a new class of large landowners. As long as he lived, he was able to keep the members of this new landed class, mainly his high civil and military officials, from developing their estates into independent centers of power. Under his successors, however, the great landowners managed to acquire a formidable position in the political life of the country.[15]

For a number of years after he was made governor of Egypt, Muhammad Ali did not have at his disposition a regularly constituted body of troops to serve as an instrument for the maintenance of order and the promotion of any ambitions abroad. The armed forces he did have were a disparate conglomeration consisting notably of the Albanian irregulars, but also Turks, Tunisians, and Algerians. As a group they lacked uniformity either in their organization or their weaponry, while they were loyal only to their particular immediate commander and to no one else. Despite their organizational and disciplinary deficiencies, Muhammad Ali still had to use them in war. In his capacity as governor of Egypt, he was assigned responsibility by the Ottoman government for overseeing matters in the Arabian peninsula. There the Wahabites, a fundamentalist sect, had by 1811 taken over a large part of the country, and it was his task to subdue them, something his forces were able to do in a series of campaigns lasting some seven years.

During his first years in Egypt, Muhammad Ali had had the opportunity to observe how efficient and effective were the British and French forces facing the Ottomans. He was certainly aware of the efforts of Selim III to create a new-style military force, the *Nizam-i Cedid*.[16] In 1815, Muhammad Ali issued a proclamation calling for the establishment of a *Nizam-i Cedid* in Egypt, although it would take him several years to work out how to organize, recruit, and officer this military force.

At first Muhammad Ali sought to utilize the Albanian irregulars as the nucleus for his new army, but as soon as he announced his intentions, they rebelled. Furious at the prospect of having to serve in a regular, disciplined force, they surged through Cairo destroying shops and venting their anger on European and Jewish businessmen. Muhammad Ali was compelled to take refuge in the Cairo citadel, fortunate to escape with his life.[17] The Albanians having proved to be unsuitable material for his new army, he thought to find a ready source of soldiers among the Nubians, inhabitants of the Sudan. Muhammad Ali thereupon set about conquering the Upper Nile Valley, in the expectation of acquiring not only recruits in abundance, but also mineral wealth. Little of the latter was found, but the Sudan did turn out to be an apparently adequate source of manpower. Some twenty thousand men were rounded up and sent to the newly organized training camp at Aswan, but they could not adapt themselves to the unfamiliar climate. Also diseases were endemic in the crowded, unsanitary conditions of the army camp with the result that at the end of six months only a few thousand remained alive.[18]

Having sought unsuccessfully to recruit two different kinds of foreign-born soldiers, Muhammad Ali turned almost of necessity to the *fellahin,* the downtrodden Egyptian peasants. In fact a few had been conscripted as early as 1820 during the campaign in the Sudan.[19] That the Egyptians might be good soldiers does not seem to have occurred to anyone for over two millenia. For all his unmilitary past, the Egyptian peasant possessed a number of attributes that could qualify him for service in a modern army—he was habituated to privation, he was sober, and he was hard-working. The peasants, however, did not accept military service willingly. In 1823 there was a revolt against the government policy of conscription. It was quickly subdued, but the following year there was a more serious outbreak, this time with religious overtones. The supression of this revolt cost the government far greater effort.[20]

Above and beyond any innate distaste the peasants may have felt towards military service, they had other justifiable grounds for opposing it. The regime was not well enough organized to conscript its forces in a systematic or equitable way. Once the government had decided how many men were needed from each district in a given year, the quota would be divided among the villages, the final choice being made by the local *shaikh.* Those who could pay a bribe escaped, while those chosen were chained together and marched off two-by-two like felons. When Muhammad Ali's power and his ambitions were at their height, the government may have been claiming as large a proportion of the populace for military service as one man out of every six.[21] Since Egypt was involved in armed conflict during much of the two decades after the organization of the *Nizam-i Cedid,* and since there was as yet no statute regulating the length of service of a conscript, he was not likely ever to return to his village.[22]

Peasants went to extreme lengths to avoid having to serve, deserting their fields, fleeing to the cities, and even leaving the country. So many had themselves mutilated through the amputation of the right index finger or the blinding of one eye that by the mid-1830s the government was threatening to condemn those involved in such acts to hard labor, which in the view, no doubt exaggerated, of the Russian

Consul would have led to three quarters of the population of the country being sent to the galleys.[23] Despite the resistance of the Egyptian populace to military service and the hardships encountered there, many observers were of the opinion that the fellahin were nevertheless better fed and better clothed than they had been before. There were fewer desertions than might have been expected.[24]

Considerably before he had resolved the issue of where and how to recruit the troops for his army, Muhammad Ali was well on the way to organizing an officer corps. An officer training course was set up at the Cairo citadel as early as 1816. In addition to taking his future officers from among the now intimidated Mamluks, since they possessed the military qualities he sought, he also continued until about 1830 to recruit Christian children from the eastern shores of the Black Sea, very much as had been done in the past.[25] It is evidence of his basically traditionalist outlook that, despite the modern implications of many of his policies, he should in effect recreate a caste of Mamluks to command his army. These men were looked upon as his personal slaves.[26]

During the first quarter-century of its existence, the majority of the officer corps were Turks. This designation was meant to include men from almost anywhere in the Ottoman Empire outside of the Arab lands, be they Turks, Albanians, Kurds, or Circassions.[27] Their common denominator was that they all spoke Turkish. For much of the 1830s Muhammad Ali's forces were engaged in open hostilities against the sultan, so Turkish officers were thus obliged to bear arms against their ruler. Under the circumstances, Muhammad Ali had to pay such men several times what they would be earning in the Ottoman forces. He also bestowed upon them grants of land and even wives from his harem.[28]

For help in the creation of an officer crops, Muhammad Ali turned to European military experts. The great majority of these were French, presumably because of France's reputation in military matters and because in the aftermath of the Napoleonic Wars there were a large number of demobilized soldiers in that country anxious to continue their military careers, even if they had to emigrate to Egypt.[29] Europeans were far more welcome in the Egyptian army than they were in Ottoman service. Still, very few of them were ever incorporated into the military hierarchy, since Muslim soldiers were as unwilling to be under the command of infidels in the Egyptian forces as in the army of the Ottoman sultan. Rather the Europeans acted as instructors and supervisors attached to various units. Among the Europeans serving with the army of Muhammad Ali was Colonel Sève. Like so many others a Napoleonic veteran, Sève held the rank of colonel only by self-acclamation, having been no more than a lieutenant, and possibly only a corporal in the French army.[30] On the basis of his talents as a trainer of men and his competence in the field, he became one of the most important figures in the realm. Sève established his ascendency over the unruly and insubordinate officer cadets at Aswan in one dramatic confrontation during his early years in Egypt. Any discontent occasioned by his being a European was stilled through his conversion to Islam.

Most of the Europeans taking service in Egypt did so as private individuals. If Muhammad Ali was glad to avail himself of the talents of these people, he also recognized the danger of becoming too dependent on them. Non-Muslims serving only for pay, they were less amenable to the kind of domination Muhammad Ali could exercise over his Muslim subjects. Then too, Europeans could function as agents for their respective governments, as happened in the case of a French mission sent out in 1824 at Muhammad Ali's request to help train future officers. The French government considered it as a means to extend its influence and perhaps even ultimately assert a measure of control over Egyptian policies. This, in any case, seems to have been the gist of the secret instructions given to General Boyer, the head of the mission. Because of a quarrel which developed among the personnel in the mission, one of them informed Muhammad Ali of the true goal of France there, with the result that Boyer and several of his collaborators had to return home.[31] Even after the equivocal attitude of the French government had been revealed, France still continued to look upon Muhammad Ali as a client and to favor him in his disputes with the Ottoman Empire. He did not reject this support.

The real debut of the new army came in the Greek revolt. As a major vassal of the sultan, Muhammad Ali was expected to contribute substantially to the Ottoman armies fighting the Greeks, and he was not yet in a strong enough position to refuse. Then, too, Muhammad Ali may have had visions of establishing himself in Greece for his own economic purposes.[32] The arrival of the Egyptian forces in 1824 under the command of his son, Ibrahim, momentarily reversed the tide of the war in favor of the Ottomans. It might well have resulted in an Ottoman victory, had not the Empire been attacked by Russia in 1828. The vigor and relative efficiency of the Egyptian forces only highlighted the decrepitude of Ottoman military institutions, graphically demonstrating to Mahmud II that drastic reform measures were indeed necessary.

By the end of the 1820s Muhammad Ali, even though still a vassal of the sultan, had acquired an extensive personal empire. Nubia and the Hejaz in the Arabian peninsula had been won by the force of arms even before he had organized his new army, while he had also been named governor of Crete as a reward for his efforts in fighting the Greek rebels. Muhammad Ali's ambitions seemed to grow in tandem with the expansion of his empire, for he now turned his sights on Syria, a land far closer to the center of Ottoman power than those acquired so far. Any gesture in that direction was sure to be regarded by the sultan as a direct challenge.

In addition to his surging ambition, a number of other factors may have led Muhammad Ali to try to acquire Syria. Historically whoever ruled in Egypt had always considered Syria to lie within his sphere of influence. Several important trade routes traversed that country, and control over them could be of considerable importance to the expanding commerce of Egypt, a matter to which Muhammad Ali was highly sensitive. Syria also possessed extensive resources, both human and material, while it could also be seen as an outlying bastion protecting Egypt against the possible plans of the Ottoman government to reestablish its control there. Muhammad Ali may also have felt impelled to act just because he had so large an army. That army constituted the chief guarantee of his power and independence. To have kept it idle could well have led to its decay and incipient dissolution; to counter that possible danger, the army "was given Syria to devour."[33] By invading Syria, the Egyptian army, heretofore victorious against primitive tribes or not very well organized adversaries, was taking on the Ottoman Empire itself, a state still assumed to be the dominant military power in the Middle East.

Throughout the Syrian campaign, the Egyptians' superiority in training, discipline, and organization brought them victory, even when outnumbered two-to-one.[34] As a result of his spectacular military successes, Muhammad Ali was made governor of Syria, to which province he then extended the harsh Egyptian system of conscription and taxation. The second Syrian War was launched by the sultan seven years later in response to Muhammad Ali's provocative declaration that his possessions were now hereditary. The sultan's hopes of undoing the outcome of the first war were dashed when, in the one great battle of the war at Nezib in June 1839, the Ottoman forces were routed.

The victory of the Egyptians in the two Syrian campaigns should not lead one to overestimate the quality of their army. Obviously it was superior to the armed forces of the Ottoman Empire, but these were still in a disorganized state following the massacre of the Janissaries and the initiation of a major program of military reform under Mahmud II. In any war with a European force the deficiencies of the Egyptians in equipment, training, and logistical support would have been patent.

With their victory at Nezib, the way was open for the Egyptians to march on Istanbul and possibly to bring about the collapse of the Ottoman dynasty and with it the disintegration of the Empire. That the European powers could not accept. Acting with rare unanimity, all of them except France, still looking upon herself as the patron of Muhammad Ali, came together in 1840 and imposed on him the Treaty of London, whereby Egypt was obliged to withdraw from her Syrian conquests and to reduce her swollen army to eighteen thousand men. In return, Muhammad Ali and his heirs were named as hereditary rulers of Egypt under Ottoman suzerainty. For all his grandiose ambitions, Muhammad Ali was a profoundly realistic person and quite ready to accept these terms.

Even without European intervention, one may question whether the Egyptian military machine would not have broken down under its own weight. An army incorporating two and perhaps three percent of the population year after year, as it did in the 1830s, would be difficult for a modern, industrial society to support. For one that was still at a relatively underdeveloped stage with respect to the majority of its institutions and mores, the burden was overwhelming, leading to the depopulation of some areas and a lessening in agricultural output as the peasants fled their land to escape the merciless exactions of the government or perished in the course of their military service. Travelers reported a mounting sense of desolation in the countryside during the later 1830s.[35] Total internal collapse might well have been the consequence of the further pursuit by Muhammad Ali of his ambitions, even if he had acquired more territories to exploit. In any case the last eight years of his reign were a period of peace and comparative tranquility.

To man, maintain, and pay for the armed forces created by Muhammad Ali necessitated the organization of a far more imposing fiscal and administrative system than had existed before. The instruments of rule of the Mamluks had generally been of a rudimentary, loosely structured nature. If sufficient to the needs of this factionalized military oligarchy, they were not adequate for the regime Muhammad Ali was in the process of constructing. During the years that he was consolidating his rule and exerting his authority over all the other politically significant elements in the country, he tended to rely on members of his own family to carry out his wishes. Muhammad Ali's system of administration was originally informal in structure, but once his power was secure he set about organizing executive departments on a more permanent basis. Each department was responsible for a specialized function of government, and each was under a designated head or minister. It was clearly modeled on the French system.

Even if the structure of the state administration was modern, it was still highly personalized in the way it functioned. All the ministers were the direct, immediate collaborators of Muhammad Ali, responsible only to him and loyal to his person and not to the bureaucratic system or to some abstract purpose of state.[36] As was the case with the army, the leading posts in the new bureaucratic machine were staffed by Turks. A ministry of war was the first permanent department of state to be instituted, about 1820.[37] Along with organizing the central administration, Muhammad Ali and his collaborators also set up a single, uniform system to oversee local affairs. Here again, the model was furnished by France. Egypt would have her own version of the Napoleonic system of prefectures and subprefectures, all with their officials carrying out the initiatives of a single authority at the center.

The tasks assigned to this new apparatus of state were extensive. Not only were the government officials to manage the workings of conscription and to collect the taxes needed to finance the ambitious policies of Muhammad Ali, they also had to oversee and control the cultivation of various crops according to the desires of the government. Imposing as may have been the structure of the state, it was also clumsy and inefficient in its operations, while the average local official was rapacious and corrupt.[38] It was through the exactions or extortions of such people that the Egyptian peasant was for the first time brought into close, regular contact with the workings of the centralized state.

The organization of a large army as well as the bureaucratic apparatus to support it required the services of men possessing new kinds of abilities. Muhammad Ali had to promote the establishment of secular schools to train men in the necessary skills. In this, the experience of Egypt under Muhammad Ali was comparable to that of Russia in the eighteenth century or of the contemporary Ottoman Empire. Since the immediate concern of Muhammad Ali, once he had consolidated his position as a ruler, was to have a body of men capable of running his armed forces, a military training school was the first one opened. As noted above, he founded such a school in 1816 at the Cairo citadel. Here some eighty men studied a variety of modern subjects.[39] The school was transferred to Aswan in 1820, a site chosen because it was distant from the capital and its perennial intrigues. Over the next fifteen years, a number of special military schools were established to train men for the different branches of the armed services as the need made itself evident—artillery, cavalry, engineers, and the navy.[40] There were also schools instituted for the training of men in less specifically military matters: a medical school, a veterinary school, and a school of languages, although the talents acquired in these places tended to be utilized above all by the armed forces.[41]

In addition to establishing military and technical schools in Egypt, Muhammad Ali sought to avail himself of the most up-to-date knowledge in these areas by sending people to Europe to study. The first educational mission to Europe was dispatched as early as 1809. In the first big mission, which consisted of forty-four people sent out in 1826, the majority were meant to study military subjects, but sixteen of them were concerned with learning about a variety of commercial and manufacturing skills.[42]

The inability of students at the special technical schools to comprehend the most basic elements in the curriculum led the government to recognize the need for some kind of preparatory course of study. Accordingly the government set about establishing a system of primary and secondary schools. There were to be secular primary schools in each of fifty provincial centers as well as two large secondary schools to prepare students specifically for entrance into the advanced establishments. At all levels of the system education was free. In the secondary schools as well as in the special technical schools, the students were subjected to rigid military discipline, while instructors held military rank in accordance with their position in the school hierarchy. At first schools were organized along the lines of army camps,[43] and until 1837 the whole educational enterprise in Egypt was under the direction of the Ministry of War.[44] It is possibly significant that the original inspector of schools in Egypt was none other than Colonel Sève.[45]

To staff his more advanced schools Muhammad Ali relied heavily on European instructors, but they generally knew no Arabic and had to teach through the medium of interpreters. It was not a satisfactory method. Because of the disadvantages here, and because Muhammad Ali was conscious of the danger of becoming too dependent on the Europeans, he tried to have them replaced by Egyptians as rapidly as possible.[46] The student body of the military schools was recruited almost exclusively from among the Turkish-speaking element; about the only opportunities for advanced training offered to the native-born Egyptians were in the Medical and Veterinary Schools.[47]

Although Egyptians may have been conscripted into the ranks of the army, the government did not for long think them fit to serve as officers. Only towards the end of Muhammad Ali's reign did it become possible for native-born Egyptians to attend the more purely military schools, perhaps because of the losses occasioned by the frequent wars. As of 1846 there were 517 indigenous Egyptians in the officer corps, but none had risen any higher than the rank of captain.[48]

On paper, the system of state-supported, comprehensive public education was a most impressive achievement. Secular education in Egypt was under a single department of state and responsive to a single central authority. In this respect it was much like the French system of education as it had been reshaped in the revolutionary and Napoleonic period. But despite the apparent rationality of its structure and the close control exercised over it by the state, the educational system did not function very well. Students tended to be selected in a haphazard manner, and once they had been trained, at no inconsiderable expense, the state did not know how best to utilize their talents. Moreover, there were never sufficient funds available to finance the schools on a regular basis. Equipment and supplies, including food, were always in short supply.[49]

The great mass of the Egyptian people did not voluntarily seek out the educational opportunities offered them by the state even when they were bestowed free of charge. In their eyes, attendance at school was the same thing as service in the army, with some youths being drafted to study, as others were drafted to become soldiers.[50] Both army and school represented a brutally sudden exposure to modes, mores, and values completely at variance with those to which they were accustomed, all in the name of goals having little relevance for them. It is hardly surprising that most of the people subjected to this hastily constituted system of modern, secular education were not, contrary to the expectations of Muhammad Ali, transformed either intellectually or socially.[51]

The elaborate structure of education established by the mid-1830s was barely in place before it began to be dismantled. A variety of difficulties beset the country at that moment, including drought, epidemic diseases, and a general economic crisis. Also, once the great struggle with the sultan was over, and his military ambitions had been curtailed by the European powers, Muhammad Ali ceased to take so intense an interest in the education of his people. As one contemporary noted: "Since he no longer had any need for an army, he did not wish to have the schools either."[52] In any case, an army of eighteen thousand men as prescribed by the Treaty of London did not require a very large officer corps. Several of the special technical schools remained in existence along with one of the two preparatory schools, but most of the primary schools disappeared.[53] The state school system would lie dormant through the reigns of Muhammad Ali's immediate successors, but it would be revived following the accession to the throne of his grandson Ismail in 1863. During the reign of Ismail the sums appropriated for education rose by a factor of ten.[54] For the first time, "the education of the people was conceived of as something to be achieved for itself alone, quite apart from any military considerations."[55] Even so, the armed forces are calculated to have absorbed some 63 percent of the graduates of the state schools between 1865 and 1875.[56]

One significant aspect of the educational policies of Muhammad Ali was the founding of a state printing press in the early 1820s. Although the French during their period of occupation had set up a press for their own use and had printed works in Arabic, this was apparently the first one established in Egypt under indigenous auspices. During the initial two decades that the government-run printing press was in operation, it produced some 240 works, of which the largest group by far consisted of military and naval manuals.[57]

In the estimation of Muhammad Ali a necessary element in the military strength of Egypt would be the ability of the country to provide its own armaments. Although the first modern weapons in the new army had to be obtained abroad, the government sought to make Egypt as self-sufficient as possible in this regard. Already by the time of the Greek campaign the various workshops in the military-industrial complex established by Muhammad Ali were capable of producing each month some fourteen hundred muskets of good quality along with gunpowder and bullets.[58] Factories were also set up to provide for the other needs of the army, namely cloth for uniforms as well as tanneries to furnish leather gear for the cavalry. Although Egypt had to continue to purchase military equipment abroad, "local production was certainly sufficient to reduce the military import bill by a large amount."[59]

Early in his reign, Muhammad Ali became convinced that Egypt possessed not merely the capability to meet its own military needs, but also the potential to become an industrial nation on a significant scale. Certainly there was cheap labor in abundance, while the state exercised a monopoly over the supply of the chief raw material, cotton. An industrial capacity would contribute to the power of the state and further assure its independence, but there existed a number of obstacles here, chief among them being the almost total lack of sources of energy. Wind and animal power were insufficient to drive a modern industrial plant.[60] So the government imported coal, along with iron, timber, and machinery, and proceeded to set up several model factories. Muhammad Ali was an obstinate person, and having decided to launch an industrial revolution on the Nile, he would not be swayed.[61] If judged in terms of the military needs of an autarkic state, the results were not negligible, the chief drawback being that most of the enterprises were operated at a steady loss.[62] The labor force for the factories was obtained by methods akin to conscription. Although a peasant preferred to go to a factory rather than serve in the army, his profoundest desire was to till his fields, even though factory work paid more.[63]

The reduction in the size of the army to eighteen thousand men meant that the products of the state arms factories were no longer so sorely needed. Then the commercial treaty between Great Britain and the Ottoman Empire, the provisions of which were extended to Egypt in 1842, undermined the other enterprises. With the tariff on imported goods set at 3 percent ad valorem, the infant industries could no longer be protected. They languished as foreign commodities flooded the Egyptian market.[64] Muhammad Ali's program of industrialization proved to be, like his system of education, a transient phenomenon, interesting as much for the energy invested in it as for the results obtained.

The Treaty of London, which deprived him of much of the extensive empire won by force of arms, would seem to have marked a period in the reign of Muhammad Ali. Considering what appeared possible at the height of his power in the mid-1830s, that reign may be seen to have ended in relative failure, but in terms of what he set out from the start to accomplish, this view has to be qualified. By the provisions of the Treaty of London, his own basic goal was achieved; his descendents would occupy the Egyptian throne for as long as the monarchy lasted. Then, too, the army, the focus of his major concern and the primary instrument for carrying out his policies, still remained in existence, although on a much-reduced scale. Among other institutions surviving the abandonment of Muhammad Ali's policy of conquest were at least some of the schools. From among the men trained in the schools founded to provide primarily for the military needs of the new Egyptian state came those who would become the "yeast of the Egyptian renaissance."[65] Finally, the basic framework of the centralized bureaucracy remained.

Later generations have seen in the internal policies of Muhammad Ali a notable example of what has come to be called *modernization*. He established a European-style army and created a centralized system of bureaucratic administration to support it, in the process destroying the independent power of the religious authorities and subjecting them, along with almost every other traditionally autonomous group in society, to the sway of the state. But for all of these radical, modernizing measures, his outward demeanor was still that of a Turk in the accepted mould. He was scrupulous in the observance of his religious obligations and in such matters as his matrimonial arrangements. If he dressed his soldiers in uniforms cut along European lines, he himself never abandoned the turban and robes. Indeed, his whole political vision had little about it that was modern.

Muhammad Ali considered Egypt to be his patrimony, almost a private preserve which could be exploited as he saw fit. He wanted that patrimony to be as prosperous as possible, and to make it so he exercised rigorous control over the resources of the country. Further, he established armed forces to protect it against interlopers. In the whole tenor of the policies he followed, Muhammad Ali may be seen as a good mercantilist, and like any mercantilist he understood that the economic and politico-military aspects of government were interrelated. Military strength would permit the government to maintain internal order, something conducive to economic prosperity, and, where possible, to enlarge the patrimony by conquest. At the same time, the wealth thereby acquired made possible the maintenance of a competent armed force. As for the Egyptian people, the inhabitants of his private preserve, Muhammad Ali looked upon them with contempt and was disdainful of their native elites.[66] They were to be treated purely and simply as subjects who should obey with no thought of participating in the business of rule. If he was interested in the modes, techniques, and gadgets of the West, it was because they aided him in the pursuit of his traditional, patrimonial goals. Muhammad Ali never had any strong feelings about whether modernization was a good or bad thing. Rather he simply saw that modernity in certain respects "presented indisputable advantages for his attaining power and establishing a strong dynasty."[67]

The military and administrative instruments created by Muhammad Ali to further his personal ambitions were more systematic and efficient than what had existed in the past. From the point of view of the Egyptian people, those characteristics also made them more oppressive than anything experienced before. But by being placed in direct contact with the source of power and authority at the center, through conscription, corvées, and other state-imposed exactions, harshly oppressive though they may have been, the fellahin were made conscious of their common membership in something larger than the village or other traditional local community. The local community may have remained the chief focal point of an Egyptian's sense of a public existence, but it was no longer the only one.

Muhammad Ali established an administrative apparatus to implement the defense of his patrimony. That apparatus had been created to carry out his will and could thus be conceived of as an extension of his personality. But as has happened elsewhere with similar organizations, this administrative apparatus began to take on a life of its own, to function as an autonomous corporate body, quite apart from the desires of the ruler, his family, and his personal entourage. In short, it began to be a state. Once created, that state by its existence, by the demands it placed upon Egyptian people, primarily for service in the army, provided the means whereby they could begin to take cognizance of themselves as a group, to sense themselves as active participants in that group and not just as passive subjects of those in power. Further, that state not only encouraged among them feelings of self-identity, but also incipient loyalty to a territorial entity. These feelings developed over several decades until they "blossomed into the Egyptian national movement in 1882."[68]

The most vigorous and militarily capable of Muhammad Ali's sons, Ibrahim was the obvious choice to succeed him as ruler of Egypt. During the last years of his father's reign, Ibrahim had already taken over a large share of the responsibility of governing, but his health was failing, and he survived Muhammad Ali by only a few months. Ibrahim was succeeded by Abbas, the grandson of Muhammad Ali but the eldest of his living direct male heirs. Abbas was the antithesis of his grandfather. Anything running counter to the explicit canons and practices of Islam was abhorrent to him, notably the Europeanizing innovations of the preceding reign. In the army recently reduced to eighteen thousand men, nine thousand were Albanian mercenaries, apparently a conscious throwback to the usages of the past and an implicit repudiation of the efforts made by Muhammad Ali to modernize his military forces. The great mass of the Egyptian people, who had borne the burden occasioned by Muhammad Ali's ambitions and their accompanying reforms, no doubt welcomed the relief brought about by the reactionary policies of Abbas. He did not wage any wars, nor build any canals, nor raise taxes. It was a happy time for the fellahin.[69] Abbas took little interest in the business of government, preferring to spend his time at his palace some distance from Cairo where he died in 1854 under rather suspicious circumstances.

Where Abbas had been of an introverted, negative character, suspicious of all and shunning the public aspects of rule, his successor Said, the youngest son of Muhammad Ali, reveled in them. He was enamoured of the ceremonial side of military life, looking on the army as a plaything, an expensive toy. The hearty, outgoing Said liked to be with his troops, drilling them and leading them in maneuvers and imaginary battles.[70] In addition to his fondness for the outward glitter of army life, Said fostered some real progress in military matters, especially with regard to conscription. The term of service was reduced to one year, and the system of recruitment was regularized, as opposed to the arbitrary procedures of Muhammad Ali's day. Said made an effort to have the sons of village *shaikhs* and other local notables serve in the army, commissioning them as junior officers after the necessary period of training. Here Said was making a conscious effort to go against the practice whereby positions in the officer corps, especially the upper ranks, were usually reserved for men of Turkish and Circassian origin. One of these Egyptian-born officers was Ahmed Urabi, the son of a village headman.

Urabi was commissioned in 1858 at the age of seventeen, and the start of his career would seem to have been especially fortunate. Named aide-de-camp to Said in 1862 at about the same time he was promoted to the rank of colonel, Urabi advanced no farther than that over the seventeen years following Said's death, primarily because with the accession to the throne of Ismail, grandson of Muhammad Ali, the Turco-Circassian element reasserted its monopoly over the positions of command within the army.[71] The circumstances of his blocked career no doubt contributed greatly to his assuming a leading role in the political-military turmoil of the years 1879–1882 that came to be known as the Urabi Rebellion.

At the origins of the train of events leading up to the Urabi Rebellion lay the extravagant financial policies pursued by the khedive Ismail.[72] He had a vision of Egypt being transformed into a modern society on the European model over a relatively short period of time through massive investments in public works, the creation of an up-to-date system of communications, and the expansion of the system of education. The Suez Canal, begun during the reign of Said but completed under Ismail, was only the most grandiose example of that vision. If it increased Egyptian prestige and catered to the vanity of the nation's leaders, the financial returns were to prove in no way commensurate with the contribution of the country in manpower and money.

The realization of Ismail's hopes may have seemed momentarily feasible with the boom in Egyptian cotton caused by the American Civil War. Following the collapse of that boom, consortia of European bankers kept the vision alive with successive loans, until Egypt had assumed a burden of debt out of proportion to the real resources of the country.

By the middle of the 1870s the government was unable to borrow any more and was in fact on the verge of bankruptcy, an unacceptable solution in the eyes of both the British and French bankers. With the backing of their governments, they were able to put pressure on Egypt to accept a joint European mission which would exercise supervision over the government's financial practices, the *Caisse de la Dette.* What was happening in Egypt with the sudden growth in the foreign debt and the consequent loss of control by the government over its own finances was similar to what would take place at about the same time in the Ottoman Empire. In accordance with the directives of the agency of the European bankers, some 60 percent of all Egyptian tax revenues were by 1877 directed to servicing the public debt. Finally two Europeans came to sit in the cabinet in order that the finances of the country be subjected to even closer oversight and control.

The policies advocated by the Europeans were classic in their simplicity: financial retrenchment and budgetary stringency. As far as they were concerned, Egypt had the resources to meet its lawful debts abroad if the government collected the taxes in an honest, efficient way and state expenditures were cut. The implementation of these financial policies led predictably to significant hardship for the Egyptian people. They were obliged to pay taxes at a higher level than before, while an army of government servants, both civil and military, saw their livelihood endangered by budget reductions.

In addition to threatening the material well-being of a large number of Egyptians, the policies being imposed upon the government raised opposition on ideological grounds. The existence of a commission of Europeans overseeing the management of Egyptian finances, as well as the more or less enforced presence of an Englishman and a Frenchman in the khedive's cabinet, represented a restriction of Egyptian sovereignty and was taken as a blatant affront by a number of groups. There was in fact beginning to emerge during the reign of Ismail a nationalist movement. Spokesmen and ideologists for the movement were the embryonic intelligentsia—teachers, journalists, and professional men of various sorts. Most of them had received their education through the school system established by Muhammad Ali with the utilitarian goal of providing trained men for the state service. The new landed aristocracy was another element in the nationalist movement. It consisted of high-ranking civilian and military officials, or their immediate descendents, whose wealth went back to Muhammad Ali's generous gifts of land. Also to be included were the rural notables, usually wealthy men of indigenous birth, as opposed to the Turco-Circassian elite. To the degree that the nationalists had a political focus, it was provided by the Assembly of Delegates, a consultative body instituted by Ismail as a means to rally support for his policies of state, but one possessing little political power. Then there was the army.

Ismail saw an imposing military establishment as a natural accoutrement of the modern, European society he was seeking to bring into existence. Having been maintained at no more than eighteen thousand men since the 1840 Treaty of London, it grew under Ismail to over eighty thousand men. So great an increase in the size of the army of necessity led to a proportionate expansion of the officer corps and the influx of men recruited from the native-born Egyptian population. By the end of Ismail's reign, they constituted a majority of the officers.[73] At the same time, Ismail, like his grandfather Muhammad Ali, apparently had no very high opinion of the military qualities of Egyptian officers. He obviously favored the Turco-Circassians. The latter monopolized the Ministry of War and filled the higher ranks, while there were only a few Egyptians who were promoted to be majors and colonels.[74] This disregard of their corporate and professional interests as well as the insult to Egyptian national dignity therein implied were infuriating to the native-born officers.

A number of other factors contributed to heightening the discontent of the soldiers. Egypt had in 1875 launched a military campaign against Abyssinia, and its unfortunate outcome was, in the opinion of the company and field-grade officers of Egyptian nationality, due to the incompetence of the Turco-Circassian high command.[75] Adding further to their discontent was the general irregularity in military pay caused by the financial difficulties of the 1870s. It was usually native-born Egyptians—the rank and file and the subaltern officers—who suffered, while the Turco-Circassians in the higher ranks were paid on schedule. Egyptian soldiers could hardly be unaware of the rising nationalist currents or immune to nationalist propaganda. This propaganda tended to take on an antikhedivial flavor, since the policies of Ismail more than any other factor had led the country to its present unfortunate state. A number of conspiratorial societies took root among the Egyptian officers for the promotion of nationalist goals. As early as 1877 one of the leading figures in this movement and spokesman for the discontent felt by his fellow native-born officers was Colonel Urabi.[76]

Among the different social groups in the process of being politically mobilized, the military occupied a crucial position. Those in the new landed aristocracy were originally of Turco-Circassian origin, and by that fact, as well as by their wealth, they were cut off from the mass of the people. As for the nationalist intelligentsia of journalists and professional men, they, like the emerging Egyptian bourgeoisie, were essentially an urban class. The military, on the other hand, could justifiably claim to be deeply rooted in the life of the fellahin. Even if recruited from the sons of village notables, rather than from the peasantry, the Egyptian officers shared many of the attitudes of the men in the ranks. Compared to the other groups in Egyptian society, the native-born officers had a high degree of unity, at least insofar as they were in general agreement about their corporate demands. This, along with their membership in a tight-knit, hierarchical organization and their cohesion, not to mention their access to the instruments of violence, gave the Egyptian officers the potential to act with a decisiveness which other social groups could not equal. In a society characterized by a still inchoate political culture and few secular institutions possessing much in the way of legitimacy, they had a degree of power and influence all out of proportion to their numbers.

A measure retiring some twenty-five hundred officers and placing them on half pay may be taken as the start of what came to be called the Urabi Rebellion. Taken by the Egyptian government in the name of economy, the measure was seen as having been instigated by the European financial commission. The officers to be retired were, of course, predominantly native-born Egyptians. To protest the government's plans, some one hundred officers accompanied by their troops and followed by a crowd of onlookers staged a demonstration outside the Ministry of Finance on February 18, 1879. In the process, the demonstrators manhandled the chief minister of the government, Nubar, and the English member of the cabinet. The two men were saved from serious harm only by timely intervention on the part of Ismail. He promised to rescind the offensive measure in question. That an opportune display of organized force could lead the government to reverse one of its political decisions was a lesson duly learned from the February 18 demonstration by Urabi and his fellow Egyptian officers.

Over the next three years, the khedivial government tried with ever-decreasing success both to regain its fiscal independence from the

European debt commission and to reestablish its authority over the country in general and the army in particular. In this last goal, the khedive was seconded by the Turco-Circassians at the head of the army, as always contemptuous of those they called "the peasant officers." At the same time, the temerity of the Egyptian-born officers, having tasted victory in an overt confrontation with the powers that be, grew accordingly. In their efforts to further their corporate goals they were much assisted by the weakening of the khedivial authority brought about when Ismail was deposed at the request of the British and French governments in June 1879. Alarmed at his efforts to escape European tutelage and to embark on policies they considered to be financially irresponsible, they prevailed upon the sultan to remove his still nominal vassal from office, a request with which Abdulhamid II was glad to comply. By having Ismail deposed and replaced by his son Tawfiq, the British and French governments did much to destroy whatever authority the khedivial office might still possess.

Tawfiq lacked the strength of character and perhaps the ruthlessness to rule in so troubled a time, but he made a valiant effort to assert his power. It was to little avail. Every attempt to oblige the army to function as an obedient, subordinate instrument of state under the control of the Turco-Circassian high command was stoutly resisted by Urabi and his military supporters. When the government tried to transfer several of them to more remote parts of the country and away from the intrigues of Cairo, there was much agitation within the army, leading finally to the dispatch by Urabi and two of his fellows of an open manifesto to the government calling for the removal of the Circassian minister of war and the appointment of an Egyptian in his place. As a result of this act of insubordination, the three colonels involved were brought before a court martial which opened on February 1, 1881. They were rescued when their own troops stormed into the room where the court was sitting and, after mishandling its members, carried the colonels out in triumph. The government, at this point incapable of responding to so overt an affront to its authority, could do nothing but dismiss the Circassian Minister of War, replacing him with the candidate of the Urabists. It was beginning to be evident that the soldiers would follow the Egyptian officers but not their Turco-Circassian colleagues, and that the "peasant" colonels with their regiments were becoming the paramount political force in the country.

Much emboldened, the Urabists staged a mass demonstration outside the palace of the khedive on September 9, 1881, and presented him with a list of their demands. These included an increase in the strength of the army and the dismissal of his chief minister. Again Tawfiq had to give in. At least one faction within the elite of the country approved of the insubordinate behavior of the Egyptian officers, the indigenous rural notables. Although they were represented in the Assembly of Delegates, their hopes of obtaining a measure of political power commensurate with their wealth were blocked by the monopoly exercised there by the Turco-Circassian elite. The indigenous notables saw an alliance with their "sons and brothers" in the officer corps as a means to achieve what they wanted.[77]

Through having successfully defied the government and made it surrender to the demands of the rebels, Urabi had become a national hero. Already the leader of the native-born officers, he was now beginning to be seen as the spokesman of the Egyptian people as well. To this role he brought a number of undeniable qualities, including an acute sensitivity to the plight of the fellahin. A plain man of peasant origin, who was endowed with considerable oratorical ability, Urabi was peculiarly able to articulate the yearnings of the Egyptian masses, stimulating their imagination and leading them "so far as to aspire to independence of both Europeans and Ottomans."[78] They came to look upon him as their true representative, calling him *al-Wahid,* "the Only One."[79] In October 1881 he publicly denounced the employment of Europeans by the government and reminded the representatives of the European powers that the army was quite capable of achieving whatever domestic goals it sought.[80] Four months later the culmination in the rise of the Urabist officers came when they imposed upon the khedive their own specific choice as prime minister, while Urabi himself was named minister of war. Faithful to his original mandate, he decreed the promotion of a large number of Egyptian officers along with a hefty increase in their rate of pay.

Urabi and his comrades had begun by defending the corporate interests of the native-born officers. In so doing they found themselves engaging first in conspiratorial and then in openly insubordinate activities. Their success here eventually led to an attempt by the government to restrain them, prompting on their part what was in effect a mutiny, when their troops rescued them from the court martial in February 1881. All the while, the Urabists were coming to see that only through overtly political actions would their corporate interests be assured. In the end the situation had so far evolved, or degenerated, that the Urabist officers took over the effective control of the government. At the same time they also assumed the leadership of a movement aiming at what amounted to a national revolution. In the process, Urabi had taken on the role of "protector of the fatherland and of Islam from the unbelieving and arrogant European powers and the liberator of the people from the tyranny of the Turco-Circassians."[81]

Because of the activities of the Urabists, the level of tumult and general disorder in the country was rising. In the face of the agitated, xenophobic state of the people, the British and French governments, out of fear for the safety of their citizens living in Egypt and also concern for their financial stake in the country, sought to induce the restoration of a measure of calm. The two governments thus sent a joint note to Egypt in May 1882, demanding that Urabi be exiled and the cabinet dismissed. It was a serious political blunder. As a result there were armed clashes in Alexandria between Christians and Muslims, causing numerous deaths. Meanwhile a British squadron had been dispatched to Alexandria. Following Egyptian rejection of an ultimatum demanding that the fortifications guarding the harbor be dismantled, the squadron opened fire. The Urabist government then declared war, calling for a general levy of the Egyptian people and decreeing the deposition of the khedive, by now virtually a ward of the British.

When the Egyptian forces finally met the British at Tel el Kebir on September 13, 1882, they were routed. The Urabist movement disintegrated, and the British occupation of Egypt began. It was to last in one form or another for the next three quarters of a century. Among the many casualties in this unfortunate affair was the Egyptian army. Armed forces were indeed to be reconstituted in Egypt after 1882, but instead of being under the domination of a Turco-Circassian caste, they were now organized and commanded by British officers mainly from the army of India. It is not to be supposed that the Egyptians found them in any way preferable to the Turco-Circassians.

Many of the factors contributing to the sad denouement of 1882 were implicit in the remarkable achievements of Muhammad Ali. His overriding aim was to make Egypt the dynastic property of his family. The pursuit of this political goal required that he be able to manage the human and material resources of the country with the least possible hindrance, and that more or less presupposed the destruction or neutralization of any group in society capable of opposing him. Muhammad Ali was able, thereby, to achieve something approaching autocratic power for himself and his descendants. Nevertheless, the situation of the ruler of Egypt, be he governor or khedive, lacked legitimacy, in that his pretensions and exactions were not necessarily accepted as inherently justified or natural by the Egyptian people in the nineteenth century. Thus the power of the ruler of Egypt depended not on any attributes intrinsic to the office itself, but rather on the vigor of the man holding the office and his willingness to use force if need be. As a result of Muhammad Ali's policies, there may have been only weak institutional restraints on the exercise of the ruler's power in Egypt, but there were also few secondary social or political reinforcements for it. If any secular institution in nineteenth-century Egypt may be considered as having over a period of decades achieved a kind of inherent authority and even legitimacy in the eyes of the people, it was the army. A sizeable portion of the young men of the country passed through its ranks, albeit unwillingly, during the reign of Muhammad Ali. Despite their resistance to military service, the fellahin were aware of the exploits of the Egyptian army against the Ottomans and took some pride in them, all of which contributed to an embryonic sense of nationalism among them. By that token and just because military service was an experience undergone by so large a proportion of the young men in the country, the army may be envisioned as a crucible in which the Egyptian nationality took shape.[82] The army was also becoming at the same time a vehicle of social advancement, as the career of Colonel Urabi demonstrated. Its notable expansion under Ismail meant that many more career opportunities were opening up to the native-born Egyptians, even if the very highest positions were reserved for the Turco-Circassians. Only the bureaucracy afforded similar opportunities to the fellahin, but compared to the army, the bureaucracy possessed less éclat and prestige.

The lack of inherent legitimacy in the khedivial office and the absence of other effective centers of social and political authority could become acutely inconvenient at a moment of crisis. In the political confusion of the mid-1870s stemming from the impending Egyptian bankruptcy, various sectors mobilized in the defense both of their own immediate interests and what they were coming to consider the national interest. The military were far better situated to assert themselves politically than were other interest groups. For one thing, the corporate interests they were defending—pay and jobs—were far easier to define, while almost by definition their reason for existence was the defense of the well-being of the country. Because the great mass of the people looked upon the military as a more disinterested body than the others, they enjoyed a widespread sense of public acceptance.

Once the military were fully mobilized politically, the problem then arose of what limits could and should be set on their activities. In the aftermath of Muhammad Ali's reforms, there now existed no subordinate institutions in society which either singly or as a group could be mobilized to reinforce the political authority of the ruler, or what little of it remained as a result of the mistakes and well-meant fecklessness of Ismail. In the Assembly of Delegates Ismail had sought to create a body capable of enlisting public support for his policies, but he had found its independent tendencies disconcerting, and in any case that body was hardly out of the embryonic stage when the great crisis over the national debt began to take shape. When Urabi moved from being simply the vigorous spokesman for the interests of the native-born officers to the more messianic role of defining and defending Egyptian national interests, no group or institution had the authority seriously to dispute his pretensions. The tragedy is that in so acting, Urabi and his followers in the officer corps were setting the stage for the attack by Great Britain and the destruction of the army as it had evolved over the past half-century, along with the abolition of whatever independence Muhammad Ali had been able to assure through his program of military reform.

As a closing parenthetical note, the Urabi episode contained one suggestive portent for the future. It would appear to be the first example of what has come to be a common feature of twentieth-century civil-military relations, both in developed and the so-called underdeveloped lands: the "revolt of the colonels." Why it should be the colonels who take the lead in confrontations with the civilian authorities and in military interventions in the political process is not certain. Possibly it is because only up through the rank of colonel can a person live more or less completely within the military milieu, secure in the illusion that modes of behavior and attitudes appropriate to the armed forces are equally applicable to civilian life, and that political problems are susceptible to military solutions. A man reaching general's rank has almost of necessity had some wider experience of the political sphere. He is less likely to retain his illusions in these matters. But the very process of having his assumptions revised concerning the nature of society outside the barracks may also deprive him of some moral authority via-à-vis dedicated, militarily more pure subordinates. A colonel, because he is apt still to maintain the basic sentiments about the extra-military world common to most soldiers, can possess a special kind of integrity in the eyes of his comrades. They are ready to follow his banner should he decide to act directly and with vigor over some public issue, intractable or ambiguous only according to a civilian way of thinking.

5 *The Reformation of the Armies of China, 1850–1912*

Permanent, if tenuous, contact between Europe and China began soon after 1500, and until the beginning of the nineteenth century, Europeans were in the position of supplicants. Denizens of the backward, far-western corner of the Eurasian land mass, they came to China to obtain the products of what was obviously a richer and more advanced civilization. The Chinese deigned to allow them limited mercantile relations under precisely defined conditions. Europeans were to conduct themselves in a manner proper to barbarians, who were bringing tribute to the Emperor in return for his benevolently bestowed gifts.

The first traders and adventurers from Europe were encountering a highly developed, complex society, one possessed of remarkable resiliency and stability. No other people had managed to sustain continuously and over so long a period of time a comparable level of cultural achievement, as manifested in their art, their philosophy, and their political and administrative institutions. Serenely, even smugly self-sufficient in almost every significant respect, the Chinese had not sought out Europe because they could see no reason for doing so.

One of the major factors in the longevity and durability of the civilization of imperial China had been the effectiveness of her military institutions. Over the two thousand years following the imposition of political unity by the Han Dynasty, soldiers played a vital role in Chinese affairs. It was they who protected the empire against the disruptive incursions of the fierce nomads from inner Asia, while they also bore ultimate responsibility for maintaining unity and a measure of tranquility within the huge, sprawling realm. As John King Fairbank has noted, paraphrasing a recent Chinese political and military leader: "Central power grew out of the sword."[1] Every dynasty, great or small, which took power, thereby assuming the "Mandate of Heaven" during this period of twenty centuries, did so by force of arms. Yet for all the crucial importance of the soldiers in assuring the existence, defense, and expansion of the Chinese realm, they were held in low esteem.

The Chinese tended to deprecate warlike exploits and the glories of battle. There was little worship or poetic celebration of heroism of the kind to be found in Western societies.[2] Rather the more pacific virtues were admired and championed, especially with respect to those governing the country. According to the dominant ideology, called here with a full awareness of the imprecision of the term *Confucianism,* the basis of good government lay in a ruler conducting himself with the utmost wisdom and propriety, thereby achieving a moral authority over his subjects so great that they would have no choice but to follow his example. With each of them acting properly and in a mode appropriate to his or her station in life, the harmony supposedly inherent in nature and by analogy present in any truly well constituted society would automatically result. A resort to force or to violence by a ruler in order to achieve his ends was of itself a tacit admission that he had failed to govern in a suitably Confucian manner. He had not set a proper example to his subjects. Under the circumstances, war was difficult to glorify, "because ideally it should never have occurred."[3]

The actual day-to-day government of the country was the responsibility of a small, constantly renewed caste of civilian bureaucrats. In the best days of the Empire they were recruited through what came close to being a system of rigorous impartial examinations in the Confucian classics. This governing elite of scholars was indispensable to a ruler whatever his origins. A measure of brutality and rude warlike vigor might be useful in winning the struggle for power, but more was required if the prize were to be enjoyed. To rule and administer the country demanded the accumulated expertise embodied in the scholar caste. So pervasive and powerful were the civilianate ideals they expounded that the founder of a dynasty, a few years removed from life in the saddle and the military camp, regularly assumed the guise of a sage ruler, giving himself over to refined delights like the practice of calligraphy. His successors would be unlikely to take the field in person. "Emperors evidently had better things to do."[4]

The Manchus, who had come to power in 1644, were the most successful of all the barbarian dynasties of conquest. They had evolved an ingenious system of rule. Manchus held the positions of ultimate power and responsibility but worked in close cooperation with Chinese officials at all levels of the bureaucracy. Even though they were vastly out-numbered by the people they had conquered, the Manchus succeeded in their efforts to maintain an ethnically and, to a degree, culturally separate identity while ruling the country according to the Chinese mode and partaking of the benefits of Chinese civilization.

The Manchus had originally been a vigorous and warlike people, with a social structure that was primarily military in nature. During the sixty-year-long reign of one of their more notable rulers towards the end of the eighteenth century, imperial China had attained its greatest territorial extent. The basic organizational element in the Manchu politico-military system was the *banner,* a permanently established body of fighting men. In the Manchu striking force which had conquered China, there had been eight banners, but their number was subsequently increased to twenty-four, eight of the new ones being recruited from among the Mongols and an equal number from the native-born Chinese. Distributed throughout the Empire at different strategic points, they served to protect the Manchu dynasty, or as it was known, the Ching dynasty, from the danger of an internal uprising, while also guarding certain perennially threatened land frontiers.[5]

The quality of the banner forces had seriously declined over the years since the Manchu conquest in the middle of the seventeenth century. Still retaining some of the traits of an army of occupation, the bannermen were meant to be maintained by the state, living apart from the indigenous population. They were forbidden to practice any trade or craft that might interfere with their soldering. Long years of peace and enforced idleness had led to a degeneration in their military qualities, while their stipend, once quite adequate, had remained unchanged during an extended period of rising prices, finally becoming little more than "an insufficient dole."[6] Inevitably many of the bannermen were driven to engage in petty trades to eke out their meager pay, while some were reduced to beggary.

In addition to the banners, there existed another, more purely Chinese force, the Army of the Green Standard. Where the banner forces were concentrated in a limited number of garrisons and were assigned no specific peacetime duties, the Army of the Green Standard was dispersed widely throughout the territory of the Empire in small units, filling a variety of police and constabulary functions.[7] The decline in the quality of the Green Standard forces was perhaps not so marked or precipitous as that of the banners, but by the end of the eighteenth century, they too had become all but militarily useless.

Having themselves won power through the sword, the Manchus were aware that the armed forces represented a potential threat to the throne. As a consequence, command and administrative functions were so organized that several centers of authority shared control over the forces in a given province.[8] This system of checks and balances may have worked to thwart the danger of rebellion against the throne on the part of some ambitious military subordinate, but it also prevented the armed forces from being mobilized with any speed in the event of an emergency. Inefficient as they were, the banners and the Army of the Green Standard were not cheap, probably costing the government more than half its yearly revenues.[9] In addition to the regular forces described above, the Ching military organization contained a third element, the militia. Where the regular forces ultimately depended in one way or another on the central government, the raising of the militia to deal with some temporary, small-scale emergency was the responsibility of the notables in a district, the so-called gentry.

Remarkably effective for some 150 years following the founding of the dynasty, the system of rule which the Manchus had instituted had begun to deteriorate by the early nineteenth century. In this respect, the Manchus were going through a process experienced by every preceding dynasty. A number of long-lived or able emperors were succeeded by weaklings and incompetents. This decline in the quality of the rulers contributed to a breakdown in the machinery of the state, for without constant oversight and a firm directing impulse imparted from the throne, corruption inevitably took root within the workings of the bureaucracy. It led to widespread popular discontent and eventually to outright rebellion.

One element in particular aggravated the seriousness of the situation for the Manchus, an unprecedented problem of overpopulation. By even the most conservative estimates, the number of inhabitants of the country doubled during the two centuries following the Manchu conquest, and it may have quadrupled, while the amount of arable land, although increasing, did not keep pace. The crisis thus created was especially acute in the already heavily peopled regions of the south. Beset with deep-seated social and economic problems and in the grip of an apparently ineluctable process of dynastic decay, the Ching government after 1850 had to face the most widespread and violent internal upheaval in the history of modern China, the Taiping Rebellion.

The Taiping Rebellion had much in common with the risings which in past centuries had heralded the downfall of a dynasty. What seemed to make the Taiping Rebellion unique, aside from its scale, and what frightened so many members of the elite, was its ideology. In the teachings of Hung Hsiu-chuan, the founder of the Taiping movement, there were, among many other things, purportedly Christian elements drawn from the Bible and from missionary tracts. The creed of the Taipings was profoundly anti-Confucian. Victory on their part might well have meant the overthrow of the reigning dynasty, and along with it the destruction of the whole existing structure of social and political power. Both for the elite of scholar-officials and for the local gentry, opposition to the Taipings was not only a profound moral duty, it was also a matter of basic self-interest.

The Taipings appeared at first to be invincible. They were inculcated with a highly effective discipline, inspired by the fervor of a utopian vision, and led in the early stages by an able strategist. Following the formal declaration of the "Heaven Kingdom of the Great Peace" at a small village in the south, the revolutionary movement swept north, gaining momentum and picking up adherents as it went. Within a short time the Taipings had captured the great cities of the Yangtze Valley, making Nanking, which fell to them in late March 1853, their capital.

In the face of the fervor of the Taiping armies, the regular forces sent out by the government of the Empire were defeated again and again. Finally the Ching authorities were compelled to look to the militia as a last resort. Raising militia forces against an outbreak of banditry or some similar disturbance was traditionally the responsibility of the gentry in a given area rather than the concern of the provincial administrators, who preferred to ignore such disorders as somehow derogatory of their stewardship. Although it might admit the necessity in certain cases of local initiatives for the maintenance of order, the central government was wary of extensive military operations being carried on in the provinces outside of its immediate control. In order to oversee the militia forces which were already being levied in Hunan province by the gentry, the government appointed the scholar-official Tseng Kuo-fan as Commissioner of Local Defense in December 1852. The son of a modest gentry family, Tseng had by persistence and sober hard work reached the highest ranks of the imperial bureaucracy, filling a number of important posts at Peking. In terms of his career to date and by his whole outlook, Tseng was a perfect example of the Confucian scholar-official, a matter of crucial importance to his leadership in the coming years.[10]

When the central government gave official sanction to the organizing of militia units in several provinces, it was still thinking in traditional terms. There would be, in effect, a multitude of local forces led and supported by the gentry, which in each district would meet the Taiping danger when and as it appeared.[11] Tseng and his collaborators recognized that the raising of militia units on the village or district level would be insufficient to counter a threat of the magnitude presented by the Taipings. A more highly militarized force, one organized on a far larger scale, would be required for the orthodox elite to be able to stem the tide.[12] Tseng set out to establish a veritable army.

Tseng intended that that army consist of a collection of local corps, the bond between them being the fact that the commander of each was known personally to him and by definition loyal. The commanders in turn were supposed to recruit and pay the men in the ranks, all of whom would come from the same neighborhood and would thus be bound by ties of local allegiance, mutual acquaintance, and family obligation. From top to bottom in his forces Tseng sought to build discipline and cohesion on the basis of these personal bonds. He stressed moral character as the primary criterion in the selection of officers, on the theory that if "a military leader did not display justice, humility, and vision, the troops would not obey him."[13] In Tseng's forces so strong was the emphasis on personal ties that if a battalion commander either died or retired, the unit he led could not be placed under the command of a new leader. Rather it had to be disbanded and replaced by a newly recruited one.[14]

In selecting officers for his forces, Tseng turned to what was, from the point of view of the army, a new source. He sought them from among young scholars of the province "who . . . had been serious in their studies and their philosophical writings."[15] Recognizing that part of the strength of the Taiping forces was to be found in their ideological fervor, Tseng sought to counter it by indoctrinating his troops with a reverence for the Chinese heritage and a respect for the traditional proprieties so obviously lacking in their foes.

To finance his army, Tseng early saw that the usual sources of revenue for the militia, namely contributions from the wealthy gentry in an area, would be inadequate, especially since he intended to pay his troops well. Although he did hold a special commission from the throne to oversee the organizing of militia forces in Hunan, for several years after he had initially launched his army Tseng was not assigned any official post in the regular bureaucratic hierarchy of the provinces which would allow him to tap the ordinary sources of revenue. He was therefore obliged from the start to resort to a variety of fiscal expedients, the most notable being a new excise tax, the *likin*. Introduced in 1853, the *likin* was essentially a transit tax on various commodities. Because it was a new impost not yet under the control of the central government, it was a flexible fiscal instrument, one that allowed Tseng independence in his conduct of military operations.[16] The throne acquiesced in these measures because under the circumstances it had no other choice, but it did not relish having its original mandate exceeded in so obvious a fashion.

Organized originally to protect Hunan against the rampaging Taipings, the army of Tseng Kuo-fan was utilized at first in a series of minor operations while it acquired the necessary discipline and perfected its training. Then in early 1854 Tseng finally acceded to the urgent entreaties of the throne and moved out of Hunan with a force of some 17,000 men to defend the threatened city of Wuchang in neighboring Hupei province. His army thereby "shed what little traces were left of a parochial militia and became a new fighting force of national stature."[17] Following the defeat of the last extant regular forces of the government in 1860, the Hunan army, now 120,000 men strong, and a few other similar forces of militia were about all that remained. In desperation the throne at last bestowed on Tseng an official position commensurate with the power he already wielded in actuality, the governor-generalship at Nanking, thus making him administrative head of the provinces which he was defending and from which he drew manpower and monetary support for his armies. Tseng was also named President of the Board of War, a post which allowed him to coordinate overall strategy against the Taipings.[18] Much weakened by internal dissension, the Taiping forces collapsed following the fall of Nanking in 1864.

The Hunan army may have been the first and most noteworthy of the militia armies raised to deal with the great crisis of the mid-nineteenth century, but it was not the only one. Among the other forces organized by scholar-officials were the Anhwei, or Huai, army of Li Hung-chang and the Army of Chu led by Tso Tsung-tang. In addition to the Taiping Rebellion the Ching government had to confront several other major uprisings. The Nien Rebellion, breaking out in 1853, covered sizeable areas in north central China, while there were two Muslim uprisings, one in the southwest and the other in the northwest. The suppression of these rebellions required years of hard fighting. It was not until the mid-1870s that internal peace and order were at long last reestablished.

In the past, militia forces had been disbanded once a given emergency was over, but that was no longer possible. Now the new militia-based regional armies took over, more or less by default, responsibility for the defense of the country, even though the regular forces, the banners and the Army of the Green Standard, remained in existence. Discredited as they may have been by their poor performance against the Taipings, they represented too well entrenched a vested interest for the much debilitated Ching dynasty to get rid of them. The assumption by the regional armies of the effective military functions of the Empire led to the atrophy of the traditional system of centralized control of the armed forces, at the head of which stood the Board of War.[19]

To bring about a breakdown in the authority exercised by the throne over the armed forces had not been the intention of Tseng Kuo-fan and his associates. Good Confucians all, the great scholar-officials sought to restore the power of the imperial government and to preserve the existing order. In their efforts to achieve that goal, they were eminently successful, but the so-called Tung-chih Restoration which they sparked was not a return to the status quo ante. By the nature of what they had had to do to save the dynasty, the statesmen of the period changed the institutional framework of the Empire, creating a new political equilibrium. Power was now shared by the throne at the center and by a number of provincial officials, each in charge of his own political and military machine. It was through the regionally based armies that military reform and modernization were undertaken during the remaining decades of the existence of the Empire.

In addition to fostering redistribution of fiscal and administrative power within the state, the regional armies also promoted a change in the patterns of recruitment into the governing elite. At its height, the Hunan army was a force of over 120,000 men. To cope with the logistical and administrative problems of an armed force of this size and to manage its sources of support, Tseng Kuo-fan developed a very competent private bureaucracy, one akin to the *mu-fu,* or staff of aides, recruited by any official to carry out his assigned functions.[20] Because the military forces created by Tseng, and later by his counterparts in other regions, did not fit into the traditional administrative framework of the state, they were not governed by the usual rigid bureaucratic rules. Service in Tseng's private bureaucracy was very attractive in that it offered tremendous scope for people of talent and initiative. Many saw it as a better or at least more agreeable way to enter public service than through the rigors of examinations in the classics. Nevertheless, so strong was the force of the traditional system of prestige and status that Tseng found it expedient to recommend his protégés for appointment to regular positions in the bureaucracy at the earliest opportunity.[21]

The methods utilized to overcome the Taipings and other rebels led to a significant change in the balance between the Manchus and the Chinese among the senior officials of the Empire. Before 1860, Manchus had predominated in the highest reaches of the state administration, but from 1860 to 1890 and beyond, it was the Chinese. During that period, 34 of the 44 men appointed governors-general were Chinese, as were also 104 out of the 117 designated to the post of governor of a province. About half of the latter had started their careers by serving in the regional militia armies, while the fact that one-fourth of them had not obtained the two higher degrees in the ladder of literary examinations is suggestive of how military service had become a shortcut to preferment.[22]

Historically domestic rebellion had often been accompanied by external aggression at the hands of the barbarians; so it was now. For a half-century before the outbreak of the Taiping Rebellion, the traders from the West who had established themselves at Canton were becoming progressively more restive at the impediments to their commercial activities erected by the imperial authorities. It has perhaps been unfortunate for the reputation of the Westerners ever since that the issue came to a head over one special commodity: opium. Chinese efforts to ban the importation of opium by the British and to destroy the stocks on hand led to the outbreak of war with England in 1839. In the Opium War, small British contingents outfitted with modern weapons easily defeated larger but poorly armed and organized Chinese forces. By the peace settlement, the Chinese conceded to the British the then-barren island of Hong Kong and a number of trading privileges.

The Opium War provided a graphic demonstration of the disparity between the military power of China and that of Western nations. In the aftermath of the war, a few men in positions of responsibility recognized that the conditions governing China's relationship with the rest of the world had been drastically changed by the rapid growth in the power of the West, most apparent from its advances in military technology and organization. Lin Tse-hsu, the official who had been originally sent to Canton to put an end to the traffic in opium and whose actions had precipitated a disastrous war, had a particular appreciation of these issues. In confidential letters to friends, he advocated that China for reasons of self-defense begin seriously to study the West and its ways, and further that the country purchase and ultimately manufacture Western-style ships and guns. As an official disgraced in the recent war, Lin did not find much of an audience for his views among the ruling elite.[23] If over the next decade several scholars began to compile works dealing with the world outside China, the majority of informed people preferred to look upon the Opium War as an event without real significance.[24] As far as they were concerned, serious threats to the security of the Empire had always come from inner Asia, and it was hard for them to become alarmed over the new barbarians poised around its remote maritime fringes.

Between 1858 and 1860, China fought and lost yet another war with the Western powers in defense of how she would conduct her commercial and diplomatic relations. Military action in this war was not kept at arm's length, so to speak, by being confined to the southern seacoast. An Anglo-French expeditionary force landed at Tientsin and marched to Peking. Having taken the city, the Europeans burned the emperor's summer place in reprisal for the execution by the Chinese of a number of European prisoners. Although the gravity of the latest challenge by the Europeans was for the moment overshadowed by the continuing danger from the Taipings, the imperial authorities who had lived through the humiliation of the Anglo-French occupation were not likely to forget it.[25] Over the ensuing decades a few officials would initiate measures in an effort to keep it from happening again. Their endeavors in this direction came to be known as *self-strengthening.*

The term *self-strengthening* was probably first used by the scholar-official Feng Kue-fen in an essay published in 1861. He defined it as the use of foreign military and manufacturing methods to defend the ethical and ideological foundations upon which China was built. He advocated seeking the help of the barbarians if necessary in order to launch the manufacture of modern weapons, going so far as to recommend changes in the examination system so that men skilled in these arts could be rewarded.[26]

However its details were understood, self-strengthening was an idea which began to arouse the interest of a few in the ruling elite. Even so impeccably orthodox an official as Tseng Kuo-fan was sensitive to the possibilities the utilization of Western-style weapons offered for the more rapid defeat of the Taipings, despite his belief that the conduct of war depended on men rather than on arms.[27] Tseng was quite willing to give foreign cannon part of the credit for the victories of the Hunan army as early as 1854, and he was very much impressed at the ease with which Anglo-French firepower repulsed the Taiping armies in front of Shanghai.[28] The Europeans were not averse to helping the Chinese in their efforts at self-strengthening, at least against the Taipings. Having at first assumed a position of neutrality with regard to the rebellion, the Europeans finally concluded that their interests would be better served by the perpetuation of the Ching dynasty than by its overthrow.

Of the leading scholar-officials, the person most closely identified with self-strengthening was Li Hung-chang, a close associate of Tseng and, as noted above, the founder of one of the regional armies of the Tung-chih period. From observing the British and French forces on a number of occasions during the Taiping Rebellion, Li had become impressed by their instruments of war and their techniques for employing them. He often wrote to Tseng during the early 1860s concerning the orderliness of the foreign armies and how this permitted them to utilize their weapons with devastating effect, in contrast to the slipshod performance of the Chinese forces.[29] Not necessarily any less convinced than Tseng of the ultimate superiority of Chinese institutions and values, Li was more pragmatic in his ideas about how to assure their defense.

A major example of the pragmatism of Li's approach to the defense of Chinese values was the Kiangnan Arsenal. Founded by Li with the support of Tseng, it was the largest of the arsenals established during the Taiping era to provide the various regional armies with weapons and ammunition. Where the other, smaller arsenals produced mainly old-style Chinese firearms, Li aspired to something more ambitious: to manufacture modern Western-style weapons and eventually to have it done by Chinese trained in the necesary skills. In an effort to accomplish that goal, he channeled a great deal of money into the Kiangnan Arsenal, funds coming from the recently instituted Shanghai customs bureau.

Under the impulsion provided by Li, the Kiangnan Arsenal grew rapidly. Already in 1866 it was reputed to vie in size with the arsenals of most powerful nations in Europe. By 1867 all the necessary tools and machinery were being manufactured within the shops of the arsenal, while a year later Chinese workmen employed there managed to build a steamship, the first of several to be produced at Kiangnan. In addition to its principal function, the Kiangnan Arsenal also took on a number of ancillary, supplementary missions of a more literary or intellectual nature. The manufacture of Western-style arms and munitions, not to mention machine tools, necessitated the translation on an extensive scale of suitable manuals.[30] That activity would continue over the next forty years, with the result that the translation department of the Kiangnan Arsenal came to play an important role in the introduction of Western science and technology into China.[31] A school for the teaching of foreign languages which had recently been established at Shanghai was also brought under the auspices of the arsenal, and its curriculum expanded to include not only foreign languages but a number of technical and scientific subjects.[32]

The Kiangnan Arsenal was the largest, most comprehensive manufacturing establishment founded in mid-nineteenth-century China, but it was by no means the only one. A navy yard established at Foochow in 1866 by Tso Tsung-tang produced some forty ships. It had associated with it a school for the training of naval officers.[33] During that same period Li organized an arsenal at Soochow some distance inland from Shanghai, while he greatly expanded the establishment already in existence at Tientsin following his appointment as governor-general of Chihli in 1870.[34] There were also a number of schools founded at various important cities for the teaching of foreign languages and the dissemination of other kinds of Western learning.

Despite the considerable sums of money and effort expended on the arsenals and navy yards, the goal of making China self-sufficient in the implements of war proved very difficult to achieve. The guns and munitions manufactured there tended to be inferior in quality and less reliable than what could be obtained abroad. It also cost more to produce them in China than to purchase them from the West. The deficiencies in the arsenals were aggravated by the positions in their staffs often being filled by people who had no special technical competence but rather a great deal of influence in particular circles.[35]

All the officials involved in establishing the arsenals realized that they would have to hire foreign technicians to direct them at the start, but they hoped eventually to replace the foreigners with trained Chinese. If some progress in this direction was indeed accomplished, it was hardly commensurate with the original aspirations. The reason is to be found in the persistence of the traditional pattern of Chinese education. Those Chinese employed in the labor force at the arsenals under the direction of foreign foremen quickly proved that they possessed the ready intelligence and manual dexterity to make admirable workers, but few from the educated stratum showed any willingness to acquire the grounding in mathematics and the physical sciences required if they were to fill the higher, more responsible posts. Talented young men from "respectable" families, to whom it was intended that the requisite knowledge be imparted in the recently founded modern schools, preferred to follow the traditional course of studies in the classics preparatory for the civil service examinations. They would continue to do so unless they were offered some incentive to deviate from the usual curriculum, and this the government was either unwilling or unable to do. For one thing, there was determined opposition in the highest ranks of the bureaucracy to the whole business of self-strengthening, while the Chinese state did not yet know how or where to employ the talents of those trained in "Western" matters. As a result the schools in question were generally frequented by superannuated individuals, failed scholars, or people simply interested in the small stipend which went along with enrollment.[36]

In the estimation of Li Hung-chang and like-minded officials, the self-strengthening movement involved far more than just the manufacture and introduction of new weaponry. Li sought to improve the defense posture of the country by reforming at least a significant portion of China's armed forces, namely the regional armies. The regular troops, the banners and the Army of the Green Standard, had apparently degenerated too far to even try to reform. Not surprisingly, the regional force to which Li was to devote the most attention during the post-Taiping generation was his own, the Huai army. The premier regional army, the one led by Tseng Kuo-fan, had on the orders of the throne been formally disbanded immediately following the suppression of the great rebellion. As a dedicated Confucian harboring no larger ambitions, Tseng was prepared to dissolve his forces if only because he was conscious of the fact that he was the most powerful man in the Empire. He wished to scotch the inevitable rumors attacking his probity and speculating about his potentially seditious designs.[37] The voluntary self-effacement of Tseng only added to the relative power and prestige of Li.

Li Hung-chang was to be the leading statesman in imperial China during the last three decades of the nineteenth century. Beginning as an associate of Tseng Kuo-fan, Li had at his behest organized and led with great skill the Anhwei, or Huai, army. He was appointed governor of Kiangsu province in the early 1860s and then promoted to the governor-generalship at Nanking before he was elevated in 1870 to the premier provincial post of all, the governor-generalship of Chihli. He was to serve there for some twenty-five years. Shortly thereafter he was also appointed Commissioner of the Northern Oceans, with responsibility for overseeing trade and diplomacy in the area, not to mention the customs revenues. That he accumulated so much power and authority, becoming the "prime minister" of China in effect, if not in actual title, may be taken as a tribute to his political skill and diplomatic acumen. But not the least significant element in Li's accession to so preponderant a position and in his remaining there for so long was his control of what for a period of two decades was the closest thing to a modern army existing in late imperial China.[38]

The situation of Li Hung-chang vis-à-vis "his" army and the state was a complicated and ambiguous one. Li's titular functions were as a provincial official, but Chihli was the capital province and was geographically exposed to foreign aggression. He therefore had also to act in the de facto capacity of an imperial official, while the Huai army came to be assigned primary responsibility for the defense of Peking. It was after all the most capable force in China. By its origins, by its patterns of recruitment, and by its original sources of financial support, the Huai army could still be considered a regional force, its officers being for the most part clients of Li. At the same time, the army had been raised under authority granted by the throne and was now financed on the basis of revenues sanctioned by the imperial government. Although Li generally recommended the men to be designated as the commanders of the army, they were officially appointed to their posts by an imperial edict. They in turn named their own subordinate officers, who were then given titles as Green Standard officers by the Board of War. Whatever their origins or their motives for entering the forces controlled by Li, the troops and officers "regarded themselves as serving the dynasty."[39]

Although Li had been the moving force behind the creation of the Huai army, he could not in time of peace exercise the same kind of semiautonomous power which he and Tseng had had with regard to their forces during the Taiping emergency. The court could at any time move Li and other provincial officials from one post to another, along with their sizeable armies. Nevertheless, Li's twenty-five year tenure as governor-general of Chihli is indicative both of his personal ascendency and of how far the effective authority of the dynasty had declined. According to the traditional rules, a senior official such as a governor should serve no more than three years in a given post. In addition, Li brought with him to Chihli the major elements of an army of some twenty-five thousand men rather than the small band of retainers which usually accompanied a senior official entering upon his new duties. The throne may have disliked having to allow a subordinate official to maintain and to overtly wield so much power of his own, but the military forces created by Li were seen as necessary to the defense of the realm.

Efforts by the throne to regain a measure of its military prerogatives by establishing armed forces capable of offsetting those controlled by Li were generally abortive or futile. In 1864 the official serving as governor-general of Chihli had set about reforming certain of the government troops in the vicinity of the capital in accordance with the methods and principles established by Tseng Kuo-fan for the Hunan army. The troops in question were also to be equipped with modern weapons. During his short tenure as governor-general of Chihli, Tseng worked to improve what had already been started. When Li took office he continued along familiar lines the work of his predecessors, but having at hand the Huai army, he simply integrated the government troops into the framework of that military organization.[40] Thus, a project sponsored by the government to reform its own troops ended by increasing the size of Li's military forces.

Li and the central government found themselves in a perpetual standoff over issues of political power and military control. The throne could not take any forceful measures to diminish the sway of Li, such as depriving him of the revenues necessary to maintain his armed forces, lest the security of the country be threatened, and its own situation be endangered in the process. At the same time, Li did little to increase his own power further for fear of provoking serious opposition. His political preeminence was solidly founded on his control of the Huai army, but to obtain the funds necessary to maintain that army he needed the backing of the throne and the good will of other provincial leaders able to tap fiscal resources larger than those produced by the relatively poor province of Chihli.[41]

Although Li, acting in his capacity as a provincial official, took the lead in sponsoring a policy of self-strengthening, the central government could not ignore the issues involved. The most powerful figure in the imperial court, the Empress Dowager Tzu-hsi, realized that something would have to be done to improve the defenses of the country, if only in the interests of preserving the dynasty. Conservative though she was, she did what she could to protect those who were advocates of self-strengthening, thereby earning the gratitude of a man like Li. She also patronized the traditionalists in the Peking bureaucracy, whose fundamental outlook she shared. They were suitably appreciative in their turn at being insulated from the contaminating effects of the new military and manufacturing projects. By being thus permitted to leave the more mundane techniques associated with self-strengthening to the private staffs of the provincial viceroys, the conservative elements in the bureaucracy were able to preserve intact their inner "essence," or *ti*.[42]

The Huai army and the other military units under the jurisdiction of Li Hung-chang may have been the most efficient and even the most "modern" forces in China, but that characterization of them must be understood in a relative sense. Although during his twenty-five years as governor-general of Chihli Li made a conscious effort to equip his forces with Western weaponry, he did not change their character in any significant way. His army was still in most respects far closer to the armed forces created by Tseng Kuo-fan than to those of the European world. The organization and the system of discipline were founded on the personalistic concepts introduced by Tseng, as was the selection of the officers. Even when the army had been more or less permanently transferred to Chihli, the individual units still bore the names of their old commanders, and command was generally passed from brother to brother. With regard to "his" army, Li's position was "that of patron, overseer, and arbitrator of a number of commanders who were from his native province and largely from his home district."[43]

Li was aware that the armed forces under his control were deficient in a number of areas, especially when compared to those of the Western powers, most notably in military training. At the time of his first observation of the European military at Shanghai during the Taiping Rebellion, he had been greatly impressed by the order and precision with which the Europeans conducted themselves on the battlefield, and over the succeeding decades Li would make a number of attempts to provide his forces with a nucleus of officers trained according to the Western manner. His first effort in that direction was to have a few men trained at West Point and Annapolis, but the project never really got started, mainly because of American hesitations over just who would be the eventual beneficiary, Li or the Chinese government.[44] In 1872 and 1876, he sent officers to Germany for schooling, while French and German officers were invited to China to train detachments in certain of the more strategically crucial provinces.[45] These small-scale and sporadic efforts to introduce more modern kinds of training do not seem to have had much effect on the overall state of the army.

The Chinese were relatively late in their efforts at establishing Western-style military academies. Not until the Sino-French War of 1884–1885 revealed how far behind the West the Chinese armed forces lagged were two military academies organized, one at Tientsin and another near Canton. Since the most urgent defense need of the country was for a more modern, efficient army, the delay here is striking, especially if one considers that schools for training future naval officers in navigation and marine engineering had been founded in conjunction with the organization of the various navy yards during the two decades since 1865. One authority has suggested that the explanation for the delay may lie in the vested interests of the army officers. No navy as such had existed prior to the middle of the nineteenth century. There was therefore no corps of traditionally trained officers whose status and livelihood appeared threatened by men schooled in the operation of modern warships.[46] Old-style army officers, on the other hand, did not welcome the appearance on the scene of men possessing a kind of knowledge likely to make them obsolete.

The original course of studies at the Tientsin military academy founded by Li Hung-chang was not very realistically conceived. Patterned after what was being done in contemporary European military schools, the curriculum comprised both theoretical subjects, like mathematics and the different sciences, as well as subjects of more practical applicability, that is, drill and operations in the infantry, cavalry, and artillery. Since most of the students were men chosen from the ranks and therefore likely to have had a limited prior education, the course of study was a very heavy one to hope to master in the one year it originally lasted, or even the two years to which it was shortly extended. A more ambitious program of professional education and training was initiated soon thereafter, whereby young men from among the sons of officials, or from otherwise respectable families, were enrolled between the ages of thirteen and sixteen for a five-year course of study. Upon completion of their work in the school they were encouraged to take the govenment civil and military examinations, in the hope that they would enter the army as a career.[47]

The other military academy was established near Canton by Chang Chih-tung, longtime governor-general at that city. A brilliant scholar of the old school, he had been confident in the ability of the Empire to resist pressure from abroad up until the Sino-French War. As governor-general in the south, Chang had been in a position, just across the border from Indochina, where he could observe at first hand the deficiencies of the Chinese forces, as well as be impressed by the effectiveness of European military methods.[48] Possibly because Chang was a more orthodox official than was Li, he from the first turned to the traditional elite in his selection of cadets for the military school, drawing them from among the degree-holding gentry or from their sons. He also kept some classical studies in the curriculum.[49]

On balance, these first two efforts at some kind of systematized, European-style military education had few reverberations within the heterogeneous Chinese armed forces. Officers graduating from the school founded by Li were supposed to be sent to the various units of the Huai army, there to serve as instructors and propagators of the modern military methods. They were not able to accomplish much, mainly because command was held by soldiers in the old-fashioned mold who were scornful of all these innovations. In sum, "a few Western-trained officers could make little impression on a vast officers' corps . . . sunk in ignorance, conservatism, and nepotism."[50]

Despite the aforementioned impediments to carrying out a program of Europeanizing military reform, many of the qualities requisite for the proper functioning of a modern army were to be found in Chinese society. The ones who emerged as the chief military figures of the era, Tseng Kuo-fan, Li Hung-chang, Tso Tsung-tang, and others, were educated men, possessing extensive experience in dealing with high-level administrative issues. For them the command of an army was simply one aspect of the overall duty of an official to maintain and, if necessary, restore order. The qualities of mind and character supposedly fostered through a traditional Chinese education were not much different from what one might hope to find in those exercising the highest commands in a European-style army. Certainly many of the greatest commanders of the past in Europe were broadly educated, often cultivated men: Gustavus Adolphus, Frederick the Great, Napoleon, and Moltke, to mention a few. With regard to the men in the ranks, the abnegation, the discipline necessary to the effective functioning of any modern, Western-style army have traditionally been distinctive characteristics of the Chinese peasant. With suitable training he could in a very short time be made into an excellent soldier—alert, intelligent, and hardy. What was chiefly lacking in the Chinese armies of the late nineteenth century was that element which constitutes the cohesive core of a regular European-style military force, the one which provides the essential, vital initiating impulse: "an officer corps capable of the routine fulfillment of the duties of modern warfare."[51] The deficiency here was akin to that found in the arsenals and the other self-strengthening projects.

An officer in a modern army is by definition a specialist, a person trained to carry out certain discrete, often mundane tasks on order. His military rank and, along with it, a degree of social and even moral stature depend on both the mastery of a particular specialty and on the ability to integrate it into the larger purposes of the armed forces. Ideally, and quite often in practice, an officer's place in the military hierarchy has nothing to do with his prior social status, depending rather on how well he performs his professional duties. In China, things were not so simple.

The idea that specialization in a given area should enable a person to rise in the social hierarchy and allow him to exercise command over others was abhorrent to the Confucian ethos. Command over other men depended in theory on moral worth, as developed through years of immersion in the classics and as demonstrated by success in the civil service examinations. Moral worth understood according to this criterion was generally more easily displayed by someone born into a family of wealth and status, since they facilitated obtaining the necessary education in the classics. Possession of a literary degree qualified a man for a position of command, but men of this kind disdained the qualities appropriate to a good, middle-grade officer in a Western-style army. As Mary Wright has suggested, Chinese society was divided into peasants who fought and an upper class whose values had nothing to do with military drill and front-line leadership.

> Sergeants and corporals frequently learned to handle their men properly, but seldom got promoted; men whose social status qualified them to be officers under the Chinese system were technically as incompetent as ever after years of modern training. The status that was a prerequisite to command produced qualities that were irreconcilable with effective command under new conditions.[52]

The basic contradiction between the realities of the Chinese social system and the necessities of a modern army was an extremely difficult one to resolve. It was not easy for the late nineteenth-century Chinese elite to call into question the principles upon which their society had been founded for centuries. Advocates of self-strengthening like Li Hung-chang may have been willing to innovate in specific areas, but they were also good Confucians. That many of them had devoted years to the study of the classics, subsequently passing literary examinations and entering the civil service, was a token of their underlying orthodoxy. Indeed it was to defend the basic framework of Confucian society that they embarked upon a program of modernizing military reform. Where the problem lay, one the statesmen in question would appear to have perceived only dimly, was that this society could be defended by these military modes and techniques in but a restricted fashion. It was something realized almost instinctively by more conservative statesmen.

As far as the conservatives were concerned, Tseng Kuo-fan had demonstrated the timeless, universal validity of Confucian principles when he used them to provide the moral basis and the organizational rationale for his "new style" Hunan army. If there was any lesson to be drawn from the great mid-century upheavals, it lay here. As has often been the case when a so-called traditional order has reached an advanced stage of dilapidation and become prey to attack, the response of many with a vested interest, be it material or psychological, in the status quo was to champion the regnant ideology at its most orthodox. That it might be irrelevant to the problems besetting the country was disregarded. Thus, in the estimation of an official such as the Grand Secretary, Wo-jen, the security of the Empire was not to be assured by adopting the "trifling arts" of the foreigners. He had never heard of any one using "mathematics to raise a nation from a state of decline or to strengthen it in time of weakness."[53] That could only be done by once again emphasizing the eternal Confucian virtues of propriety and righteousness.

Recognizing that a reaffirmation of Confucianism, even if undertaken with great sincerity, was no longer adequate to save the country, Li Hung-chang and his collaborators sought to be innovative and adventurous by introducing Western weapons and military technology. These would in their estimation serve as a kind of protective armor for the still viable corpus of Chinese civilization. The demonstrated ability of the Chinese to produce acceptable modern arms at the various arsenals could be taken as providing some validation for their view. But they did not foresee how difficult would be the problem of reorganizing the armed forces so that the modern arms could be utilized to the greatest effect. Nor did they recognize the full implications of that undertaking, namely the likelihood of a modernized army with a competent officer corps disrupting "the very social order the new arms were designed to protect."[54] It is not certain how enthusiastically Li would have advocated military reform if he had realized that it either presupposed or might lead to the transformation of Chinese society. The point of view expressed in the pungently anachronistic remarks of Wo-jen had at least the virtue of consistency (if only that)—something lacking in the thinking of the advocates of self-strengthening.

The climax to China's nineteenth-century effort at self-strengthening came in the Sino-Japanese War of 1894–1895. Like previous armed confrontations with Western or Westernized powers since the middle of the century, it ended with the defeat of the Chinese forces. Only a small portion of the available military manpower in the country was ever mobilized or committed to battle. Poor transportation facilities, along with provincial jealousies and bureaucratic rivalries of one kind or another, made it in effect a war between the newly modernized forces of the Japanese Empire and the armies of China's northern regions.[55] Of the Chinese forces actually engaged in combat, only the fifty thousand directly under the control of Li Hung-chang could be considered "modern" in terms of their training or their organization.[56] To be beaten by the British and the French was galling enough to Chinese pride, but so rapid a defeat at the hands of the upstart Japanese was utterly humiliating. China was obliged to concede to Japan the island of Formosa and to pay a large indemnity. The defeat also led to the downfall of Li Hung-chang, even though he had opposed China's going to war with Japan over their conflicting interests in Korea.[57]

The disgrace of Li did not mean the termination of the self-strengthening movement, but rather an extended, often confused effort to redefine what it really entailed and what it might accomplish. Where Li had thought of self-strengthening primarily in terms of improving the country's armed forces and of establishing specifically defense-related industries, a recognition was beginning to emerge among some of the prescient members of the elite that a more wide-ranging reform of the governing institutions might be necessary. Concerning the nature and the magnitude of what should be done, there was, however, in the years following the Sino-Japanese War no real consensus among the responsible officials or between them and the Manchu rulers.

For a moment during the summer of 1898, it seemed as if the central government and the dynasty might generate within themselves the energy to undertake an extensive program of reform. The effort was inspired by the brilliant scholar, Kang Yu-wei. He had made a radical reinterpretation of the Confucian classics in order to demonstrate how they were compatible with the Westernization or modernization of major aspects of Chinese social and political life. His teachings came to the attention of the young, impressionable emperor, Kuang-hsi. Under the influence of Kang's ideas and possibly at his instigation, the emperor launched the famous One Hundred Days between June and September 1898, when some forty to fifty reform decrees dealing with education, government administration, industry, and international cultural exchanges were issued with precipitous rapidity. Radical initiatives of this nature alarmed conservative officials, so much so that the upshot of the whirlwind of reform was a coup d'état fomented by the Empress Dowager. Emerging from a decade of retirement, she openly took up the reins of power she had been manipulating surreptitiously. Kang fled the country, while most of those in the reform clique were either banished or executed.[58] In the reform edicts issued during the One Hundred Days, little stress was placed on military matters.[59] On the evidence, no one seemed overly anxious about the continuing disorganization of the Chinese armed forces, which is striking if only because a crucial element in all thinking on reform, be it conservative or radical, was the growing apprehension within the ruling elite over the security and the integrity of the country. That apprehension was not ill founded.

After the stunning victory of Japan in the 1894–1895 war, the European states with interests actual or potential in the Far East had come together to make her limit some of her more extensive territorial demands. The states in question realized that they had better take steps to protect their interests against the danger implicit in the presence so close to China of a militarily strong, ambitious Japanese Empire. There followed what came to be known as the "scramble for concessions," as each state sought to assure for itself exclusive commercial rights in a number of specified ports, along with a sphere of influence in the hinterland, an area generally comprising a province or two, where the European state would have monopoly control over such matters as the conduct of trade and the development of railroads and mining. There was little the Chinese government could do about the demands of the foreign powers for the awarding of concessions, given the general military disarray of the country.

If advocacy of reform may be understood as the response of many in the elite of scholar-officials to what appeared to be the piecemeal subjugation of the country, among the great mass of the people a more primitive reaction was taking form. It manifested itself most notably in the widespread but uncoordinated development of the secret society, the Righteous and Harmonious Fists. The Boxers, as they came to be popularly called, were pledged to rid the country of all foreigners, along with their Chinese associates and collaborators. The mounting tide of xenophobic sentiment reached a climax in early June 1900, when the Boxers attacked the foreign legations in Peking. Then on June 21, the Manchu government declared war on the foreign powers. Some of the more extreme conservatives close to the throne saw in the sponsorship of the Boxers a way to build support for the dynasty within the country, and they were able to win the Empress Dowager to their point of view. Against the disorganized, if fanatical, forces of the Boxers, the small contingents dispatched by the foreign governments had no great difficulty, completely defeating them and imposing on China an indemnity of 450 million taels, or some $330 million.

Coming so soon after the Sino-Japanese War, the catastrophic outcome of the Boxer Rebellion provided a stark demonstration of how desperate had become the situation of the country. Even the most intractable and obscurantist of those at court had to recognize that the dynasty might not survive unless it could carry out a far-reaching renovation of the major public institutions. In particular, the Empress Dowager became a determined advocate of reform. During the remaining half-dozen years of her life she would be the patron of what came to be called the Manchu reform movement. Many of the measures initiated during the One Hundred Days and then aborted by the coup d'état were once again undertaken, but now in a systematic way. Over the next decade they began to give state and society what amounted to a new physiognomy. Of all the government-sponsored efforts at reform, few were of greater significance than those having to do with the army.

The first steps in this latest effort at military reform came in the summer of 1901 when two of the governors-general, Chang Chih-tung at Wuhan and Liu Kun-i at Nanking, published a number of memorials to the throne. They advocated, among other things, the termination of the traditional military examinations, the establishment of modern military schools, and the adoption of Western training methods. They also called for the abolition of the Army of the Green Standard and the institution of constructive roles for the bannermen.[60] Within a few months the throne had issued edicts setting in motion the implementation of most of these measures. It is to be noted that although the dynasty thereby showed itself amenable to a general program of military reform, the initiating impulse still came, as it had for the past half-century, from provincial officials. Only towards the end of the decade would the throne begin to assume a more overt role here.

Two figures predominated in the post-1900 reform of the armed forces. One of them, Chang Chih-tung, had been a leading provincial official for some two decades. As noted above, he had come to see the vital importance of military affairs at the time of the Sino-French War of 1884–1885. Chang was acting governor-general at Nanking during the war with Japan when, as a defense measure, he organized a force of some thirteen battalions known as the Self-Strengthening Army. Modeled very closely on the German tables of organization, this army was trained and commanded by some thirty-five German officers and NCOs. They were to serve until the Chinese could acquire sufficient knowledge and experience.[61] To place Chinese soldiers in the ranks directly under Europeans represented a major innovation and may possibly be taken as an indication of Chang's assessment of the gravity of the actual situation. When he returned to his regular post of governor-general at Wuhan shortly thereafter, he did not, in the manner of other nineteenth-century regional leaders, takes his troops with him. Leaving them in place, he set about organizing yet another local army.[62]

Where Chang Chih-tung was a scholar-official in the traditional mode, the other major military figure of the early twentieth century, Yuan Shih-kai, was not. Having failed in his efforts to pass the civil-service examinations, Yuan had had to purchase a scholarly degree and had thereafter entered the Huai army of Li Hung-chang. His undeniable talents eventually brought him to the attention of Li, whose protégé he became. In December 1895 he was placed in command of the forces hurriedly organized earlier that year in Chihli province at the behest of the central authorities. Yuan's already meteoric career was advanced yet further when in 1901 he was made governor-general of Chihli. This can be seen as tribute to his remarkable work in the organization and training of the forces in the North and to his political skill in bringing them through the Boxer Rebellion unscathed. Of course, his reputed betrayal of the reformers at the climax of the One Hundred Days in 1898 may also have contributed to his rapid promotion. For as long as she lived, the Empress Dowager would be his patron and ally at court.

In many respects the situation of Yuan Shih-kai as governor-general of Chihli appeared similar to that of Li Hung-chang in the same post. Both men owed their elevation to this supreme post in the provincial administration largely to their demonstrated ability to control considerable military forces, and both also assumed ad hoc responsibility for a number of imperial functions as well, in particular the defense of the capital. There was however, at least one notable difference between the situations of the two men. Li had himself created the military forces he commanded, developing and overseeing their sources of revenue, as had also Chang Chih-tung. Yuan, on the other hand, was originally appointed to the command of an already existing force, one organized under the auspices of the throne and supported by revenues from the central government. Lacking a comparatively independent regional base in Chihli or elsewhere, he was obliged to pay constant attention to the strength of his political situation in Peking.[63]

Since Yuan's forces had the mission of protecting Peking, and since the focus of the increasing tensions between Russia and Japan lay near at hand, he could count on the largesse of the throne. The growth of the forces under Chang Chih-tung, on the other hand, was limited by their being supported solely on the revenues of the provinces confided to his care. As a consequence, where the two armies were roughly comparable in size from 1895 to about 1904, Yuan's forces thereafter were rapidly enlarged. By the end of his term as governor-general, the northern, or Peiyang, army as it had come to be known, totalled some one-hundred-thousand men reportedly "all capable of serving in modern war."[64] Yuan introduced European-style training and tables of organization into the Peiyang army along with advanced European notions like large-unit maneuvers. If the maneuvers held in late 1905 bordered on the chaotic, those carried out the following year made a favorable impression on foreign observers.[65]

Presiding over a military force of the magnitude of the Peiyang army, Yuan Shih-kai was probably the most powerful figure in China during the early years of twentieth century. Even though he was, as far as can be gathered, a loyal servant of the throne, he was extremely ambitious, and well aware of the importance of these military forces in protecting his situation. Yuan was also conscious of the political value of having a satisfied band of protégés. He was thus assiduous in his efforts to assure rewarding careers for the young officers being produced within his forces. As there were not enough billets in Chihli to meet their legitimate aspirations, Yuan used his influence to find them positions in the armies of other provinces and then had them posted back to Chihli on a theoretically temporary basis. He thereby retained their services and built up a loyal clientele. Since the new forces were not part of the official hierarchy, Yuan and the other provincial leaders had to obtain ranks for their officers in the traditional military bodies, moribund as they were, and in the recognized branches of the bureaucracy. In addition to those possessing rank in, but not serving with, the Army of the Green Standard and the various banners, there were to be found in Yuan's forces men with the title of "expectant intendant and prefect of the civil service."[66]

A major goal of the Manchus in the early years of the twentieth century was to strengthen the situation of the dynasty within Chinese society. They therefore took great interest in improving the state of the army, while endeavoring to establish their mastery over it. The throne encouraged the efforts of Yuan Shih-kai to create an effective military force in Chihli, providing him with the necessary financial support, since that force was looked upon by the government as more an imperial than a provincial army. Yet if the government found it expedient to further Yuan's projects, the power thereby placed at his disposal represented an implicit standing threat to the long-term interests of the dynasty. No more than in the case of Tseng Kuo-fan or Li Hung-chang did the throne relish the concentration of so much military power in the hands of a subordinate official, even if there was no overt reason to doubt his loyalty.

In order to increase its authority over the armed forces, and also to impart a more uniform thrust to the renovation of Chinese military power, the government ordered the establishment of the Army Reorganization Commission in December 1903. Despite the commission's being under the guidance and domination of Yuan, one of the chief, though unavowed, goals of the government in reorganizing the army was the eventual reduction of his power.[67] To the degree that the commission sought to promote unity within the Chinese military establishment, its immediate success was meager. At the imperial German maneuvers of 1905, China was represented not by a single delegation of observers, but three: one group of officers sent by Yuan Shih-kai, another by Chang Chih-tung, and a third by the Army Reorganization Commission.[68] Here was graphic evidence of the continuing fragmented state of the armed forces.

From the deliberations of the Army Reorganization Commission, which began during the course of 1904, there emerged a series of plans and regulations governing the development of the *Lu-Chun*, or "New Army." The aspirations of the commission were nothing if not ambitious, for the New Army was supposed by 1922 to consist of thirty-six divisions, all organized according to the latest principles in honor among European armies, with a large cadre of regular troops backed by first- and second-class reserves. In addition to its grandiose general plans, the commission also produced highly detailed regulations concerning pay, training, tactical formations, and so forth. These should perhaps be looked upon as pious hpoes, since the government was far from possessing the fiscal resources or the administrative expertise to implement them.

The projects of the Army Reorganization Commission did not necessarily meet with universal assent. Chang Chih-tung, for one, hardly appreciated being told by some brand-new agency in Peking how the troops under his jurisdiction should be organized and trained. He did after all have extensive pragmatic experience in this regard. Any objections he might make to the incursion by the central authorities into his military bailiwick were circumvented by his being transferred to the Grand Council at Peking in 1907 and thus removed from his seat of military and political power. He died two years later. Yuan, on the other hand, had little difficulty in conforming to the regulations established by the commission, since he had had a large role in drawing them up, and since his forces were a major component of the New Army.[69] But he too would see his semiindependent military authority dissolve. In 1906 four of the six divisions in the Peiyang army were removed from his jurisdiction and placed under the direct authority of the Ministry of War, the name recently assumed by the Army Reorganization Commission.

A far more serious blow to Yuan's politico-military situation came some two years later with the death of the Empress Dowager, for he thereby lost his most valuable ally in the highest councils of the government. His own more or less forced retirement followed shortly thereafter. That the throne dared to dismiss an official of Yuan's prominence, one controlling the most powerful military force in the country, is suggestive of the growing self-confidence of the Manchu leadership.[70] As for Yuan himself, he accepted his dismissal without demur in the manner of a good Confucian official, even if he lacked the academic credentials of one. His dismissal did not necessarily mean the end of his influence within the Peiyang army. Most of the high-ranking officers in the north owed their careers to Yuan and were to remain loyal to him even though he was out of office.

The efforts of the dynasty to establish its control over the armed forces were hampered by a certain superficiality in its conception of how to go about the task. When in 1909 the throne published a decree solemnly proclaiming the emperor to be the commander-in-chief of the army, the measure might have been more striking had not the emperor in question been a child of three and the obvious intent of the decree to heighten the power of the regent at the expense of the provincial officials.[71] If anything, the result was less to increase the authority of the dynasty vis-à-vis the armed forces than to arouse the disdain of its Chinese subjects.

Recalling their origin as a warrior people who ruled China by right of conquest, the Manchus sought to take the lead in the renovation of the armed forces already under way. They saw themselves as a military ruling caste and believed that Germany provided the best model for their undertaking. One example of their efforts here was the establishment of an Imperial Guards Corps of Manchus, the members of which were to receive higher pay and better opportunities for promotion than other soldiers. Military reform thus envisioned amounted to little more than exchanging Confucian roles for German epaulets.[72]

The dynasty may have sought to have the army function as a uniform force under its direct command, but its aspirations here were vitiated by the persistence of the old system of administration and taxation. So long as a senior provincial official continued to discharge the essential financial functions for the troops in his area of jurisdiction, the throne could not exercise centralized control over the army. The initiatives necessary to bring about significant changes here appeared to be beyond the capacity of the Manchu rulers.

For all its shortcomings, the reform of the army would appear to have been one of the more successful undertakings in the Manchu reform movement. In its training, equipment, and morale, the Chinese army within a relatively short period of time had been almost transformed. "China's military prowess came as a revelation to observers with a knowledge of the defects of China's demoralized and underequipped armies during the nineteenth century."[73] Possibly the crucial element in the military renovation of China was the system of education instituted, especially for the officer corps. In this area too, the leadership was taken by Yuan Shih-kai.

When Yuan first set about bringing his army up to modern Western standards, he employed Japanese advisers as well as instructors from the foreign-trained troops of Chang Chih-tung.[74] Within a relatively short time, however, he was able to provide trained men in increasing numbers from his own schools. Included among the military training schools maintained by the Peiyang army were a staff college with a one-year course, a regular military academy with a two-year course, a three-year school for noncommissioned officers, and a school where old-style officers might pick up a few Western techniques.[75] The most thorough and well-planned schooling along Western lines was offered by the military academy at Tientsin, with its twelve-year course divided into three stages. By 1906 nearly one-thousand officers and N.C.O.s from Chihli as well as from other provinces had received training at one of Yuan's schools.[76] Chang Chih-tung also created a system of military education of the same order of efficacy, if not magnitude, as did Yuan. Faithful to his origins, Chang consciously sought to blend a modern military education with more traditional concepts. To a greater degree than Yuan, he emphasized the importance of literacy among his troops, retaining some elements of the classical curriculum in the course of studies.[77]

The military schools sponsored by Yuan Shih-kai and Chang Chih-tung were by no means the only ones established in China during the decade after 1901, although they were certainly the best. Similar schools sprang up in virtually all the provinces. By 1908, some seven thousand students were attending officers' schools while twenty-five hundred were in the schools for N.C.O.s, not to mention seven hundred studying in Japan and smaller numbers in the United States and Europe. The end of the decade saw some seventy military schools of one kind or another in existence throughout China.[78]

The military schools were to have a great influence on the political and social evolution of the country during the last years of the Empire and beyond. They were perhaps the most prominent sector among the newly established modern schools, where persons aspiring to positions of leadership, status, and prestige within Chinese society now had to acquire their education. In 1905, the time-honored mode of entry into the elite had been closed with the termination of the examinations in the classics. These examinations had been the focus of the whole system of education. Their abolition, at the very least, denoted a lessening in the importance of the classics and with it a change in the nature of the governing elite.

Learning still continued to furnish the major path of entry into the elite, but now the goal and content of that learning were different. Old-style officials had been broadly educated generalists, ideally more interested in achieving through years of study the moral qualities necessary for governing their fellow men than in the mere acquisition of utilitarian knowledge. By contrast, the men enrolled in the modern schools looked upon the mastery of a specified body of useful learning as the basic purpose of their education. That the new elite should consist of narrowly trained specialists, very few of them steeped in ethical and philosophical issues through long exposure to the classics, would have seemed immoral to a true Confucian. How much more distressing was the idea that many of the new leaders of the country now were men taught to be managers of violence. About the only solace to be derived from this state of affairs was the fact that most of the cadets at the military schools tended to be recruited from among the sons of "provincial gentlemen."[79] A career in the army was indeed becoming a highly respectable one as "young China went to school in uniform and young men from good families did not hesitate to enter military service."[80] As opposed to the men who had served under Tseng Kuo-fan or Li Hung-chang, the new caste of school-trained officers considered a military career as an end in itself and not simply a short-cut to a place in the bureaucracy.[81]

More than the other members of the new educated class being produced by schools established under the Manchu reform movement, the young army officers were acquiring a "modern," or Western, outlook. The uniforms they wore made them familiar with Western-style dress on an everyday basis, while their study of how modern weaponry worked and how it could be utilized most effectively initiated them into a familiarity with Western technology. In the process of being trained along Western lines at the military schools, the young officers were also exposed to philosophical and political ideas coming from the West. It was almost impossible to keep such matters from insinuating themselves into the course of study along with more purely professional concerns. Where the old elite of scholar-officials had through their education been imbued with the Confucian ideology and, by extension, a commitment to the social and political institutions embodying that ideology, the new military specialists could find themselves increasingly estranged from the existing system, largely because of the form and content of their education.

The newly awakened interest in things military on the part of so many in the elite was closely related to the growth of Chinese nationalism. Although the Chinese may be considered by any number of indices to have constituted a nation, before the last decades of the nineteenth century they were not conscious of themselves as such, nor were they ready to place an especial value on the fact of their nationhood. They were therefore not susceptible to nationalism. What triggered the emergence of this modern, powerful sociopolitical force was a rapidly growing awareness among the elite of how vulnerable China was to the depredations of outsiders. If in the past the country had on occasion been conquered, the conqueror had almost invariably adopted Chinese ways. China, then, had won in the end—proof, as far as the elite was concerned, of the superiority of her culture and of the universal validity of her civilization.

The interlopers from the West were disconcertingly different. In addition to their rude vigor, typical of all barbarians, they knew how to apply their energies in an extremely efficient way, being able thereby to overcome the far more numerous forces opposing them and to impose their commercial and political conditions on the Chinese government. Unlike other barbarians, they appeared oblivious to the superiority of Chinese civilization, when they did not hold it in contempt. Rather than seeking to partake of its benefits, they preferred to establish small enclaves within the country, where in parochial fashion they set about recreating and perpetuating their own outlandish style of life. This lack of interest in China except as a land to be exploited commercially or as a people to be evangelized according to the precepts of a seemingly crude, irrational religion, combined with their ability to penetrate into the country at will, made the late nineteenth-century confrontation with the Western barbarians an experience of a totally unprecedented kind. It was also an intensely rankling one for those in the elite, obliging them to reassess their serene, time-honored assumptions about China's place in the world.

From being the unquestioned center of civilization, a land whose traits imposed themselves on all comers through their intrinsic worth, their self-evident rightness, China was becoming a beleaguered fortress under attack. Qualities and attitudes peculiar to China had once been championed because they were believed to be universally valid and therefore superior. Now they were being desperately defended just because they were Chinese and therefore familiar. When that happened the members of the elite had become nationalists.

In the context of the existing social and political situation, nationalism had antidynastic, even revolutionary implications. Not only was the dynasty of non-Chinese origin, it also seemed to be too degenerate to lead in the defense of the nation against the danger from abroad. Further, the institutions through which China had been ruled for centuries were proving to have little utility or relevance given the problems facing the country, as these were perceived by the new elite. That those institutions of rule might have to be transformed, even abolished, was ceasing to be an unthinkable prospect.

The military were a conspicuous element in the nationalistic movement growing out of the awareness on the part of the Chinese as to how vulnerable their country was to outside aggression. The armed forces were, after all, the instrument through which the Chinese people would have to defend themselves and their land. Quite conscious of the centrality of their role, soldiers were ardent exponents of nationalist beliefs. Some of them became revolutionaries in the process, their whole "modern" frame of mind—fostered by their education and by the demands of their profession—being incompatible with the regnant ideology.

The soldiers were not the instigators of the nationalistic, antidynastic movement of the two decades after 1890. They were, however, among the favorite targets for the efforts of the early nationalist agitators and propagandists. Of all those in the army, the ones most likely to develop revolutionary sentiments were the officers trained in Japan. Here they were likely to encounter other Chinese exiled for their seditious activities. They were, in fact, sought out by the exiles on the assumption that they would be returning to China soon and thus be in a position to carry and also to act upon the revolutionary message being disseminated by the exiles. The revolutionaries understood that in any effort to overthrow the Manchu dynasty, the attitude and behavior of the armed forces would be of vital importance. Since those trained in Japan were generally assigned posts of command or important staff jobs, they were especially worth cultivating. Moreover, the young officers studying in Japan were responsive to the ideas spread by the revolutionary exiles if only because they could see all around them a country which had apparently rejected traditional ways and was making the transition to modernity with ease, gaining power and prestige on the world scene. As they pondered the problem of why China was not doing the same thing, it was all too easy to conclude that the problem lay in the Chinese political system and the decay of the ruling dynasty.

Quite apart from the revolutionary doctrines advocated by the followers of Sun Yat-sen or the modern outlook introduced to students at the military schools, a number of other factors contributed to the disaffection of the soldiers. Conditions of service continued to be poor in the armed forces, while serious flaws in Chinese military life persisted, if only because the ambitious reform programs undertaken by the Manchu government in the post-Boxer decade far outran the available financial resources. So long as the Ching state was unable or unwilling to reshape its fiscal system, it could not raise sums adequate to support the newly renovated and enlarged armed forces. Then too practices like nepotism and favoritism toward men from one's own locale continued to exist. For the new kind of recruit, entering the army out of a sense of patriotic idealism and devotion to the national cause, the persistence of old-fashioned ways in the armed forces was disheartening.

Another element of discontent within the ranks of the army arose from its never really fulfilling its seeming early promise of being an easy avenue for social mobility. Young men of ability, who possessed only modest means, found that those with money were able to acquire extensive preliminary training in the new schools and as a consequence to have easier, more successful careers than they. There was thus a growing cohort of able, but frustrated men open to the message being expounded by various revolutionary agitators, if not necessarily for political reasons.[82]

Soldiers in the garrisons of south and central China tended to be more disaffected than those in the north for several reasons. The units of the Hupeh and Hunan armies were usually stationed in urban areas, especially in the three cities of Wuhan—Wuchang, Hankow, and Hangyang—while the garrisons of the Peiyang army were more widely dispersed. Dissatisfaction over government policies and the hardship caused by pay cuts, combined with the rampant inflation of the period, led to a politicization of the Hupeh and Hunan units. Politicization was also "the natural consequence of concentrating a large body of literate young men in a volatile urban center."[83] The high degree of literacy in these units was legacy of Chang Chih-tung. He had assumed that literate men would be more effective soldiers and that by receiving some training in the classics they would be more likely to be loyal to traditional norms and ideals. But their literacy could in fact have the opposite effect, by making them more open to revolutionary propaganda.

How widespread was the discontent in the army may be gauged by the number of mutinies and other outbreaks of violence occurring between 1908 and 1911. Whether they were triggered by anger over material conditions or rather by political factors, they tended to take place more frequently in the southern units than in the Peiyang army. The troops in the north were not immune to sentiments of alienation, but the influence of Yuan Shih-kai remained strong in the area even after his forced retirement. Yuan was a reformer, not a revolutionary, and his protégés, still in command of the Peiyang troops, followed his lead, maintaining an aura of at least formal obedience to the dynasty. Furthermore, the Peiyang officers had been trained predominantly in the indigenous military schools where a sense of obligation to Yuan and to the existing order was inculcated. In the Hupeh and Hunan units, now known as the Eighth Division, there were officers in relatively large numbers who had been trained in Japan and who were thus more likely to be adherents of the revolutionary cause than their comrades in the north.[84] So serious was the disaffection among the troops that the central China branch of the Revolutionary Party and the Eighth Division had almost become synonymous. Hupeh soldiers "were indeed the backbone of all the local revolutionary bodies."[85]

When revolution came during the second week of October 1911, it was more the result of a series of fortuitous events than of any systematically laid out plot. It began with the accidental explosion of a bomb in the headquarters of the revolutionary organization at Hankow. Fearful that their conspiratorial activities would be uncovered and they themselves punished, the revolutionaries, in particular two battalions of soldiers, staged a hastily planned, almost impromptu uprising. The governor-general and the general in command of the garrison at nearby Wuchang, taken aback by the suddenness of the mutiny, fled the city, leaving the troops who remained there without leadership or guidance. They all went over to the revolutionaries. In what amounted to a chain reaction, the troops in one garrison after another, first in the Hupeh and Hunan area and then elsewhere, followed suit until by the end of the year every military unit in China outside of those under the control of the Peiyang army had joined the revolutionary cause.

Within weeks of the Wuhan uprising, the government had to undergo the humiliation of recalling from retirement Yuan Shih-kai and bestowing upon him the mandate for suppressing the revolution.

Yuan refused to accede to the requests of the throne until the government met his terms. On November 1 he was named prime minister. Although he was against the revolution, Yuan was astute enough to recognize how far the Chinese people had been estranged from the dynasty, and he saw the futility of a prolonged struggle waged in its defense. Should he win that struggle, the victory would redound above all to the credit of the Manchus, while a hard-fought civil war might harm the well-being of the Peiyang army, the force which he had so carefully developed and which was the chief basis of his power and influence.

By moving slowly and prudently, Yuan was able to nurture his own political fortunes, taking full advantage of his having become for the dynasty the indispensable man. A few months of desultory and inconclusive fighting between the imperial and revolutionary forces ended when the leading officers of the Peiyang army announced that they were no longer willing to fight for the Manchu cause. With the desertion of the northern troops, the dynasty lost the last vestiges of control over the armed forces. The only possible course was now the abdication of the Emperor. It was followed by the publication of a decree calling for the establishment of a republican regime. How large a role Yuan played in promoting the disaffection of the Peiyang generals is not certain, but he was the one who profited the most from it. On January 1, 1912, Sun Yat-sen was made provisional president of the Republic, but quickly recognizing the difficulty of the present situation and the need for someone capable of managing the existing military forces in order to maintain the precarious unity of the country, he stepped aside in favor of Yuan. Named provisional president in February 1912, Yuan aimed at manipulating the office in such a way as to assure the restoration of the Empire with himself as ruler. For a variety of reasons his efforts in this direction failed, and Yuan died in 1916, a bitterly disappointed man.

With the fall of the Ching dynasty the remains of the traditional system of government broke down as well. Only the merest pretense at centralized direction was made as the different regions and provinces tended to establish themselves as separate republics. No force appeared capable of assuming effective political and administrative responsibility over the country during this period of generalized disintegration except the military. The other sectors of the new elite lacked the strength, the cohesion, and the self-confidence for the task.

During the ensuing decades various factions would seek to reestablish a degree of unity within the politically fragmented land. Although civil leadership did not completely disappear, the real touchstone of political power was military strength. Regionally based military forces had been an important new factor in the political life of China following the Taiping Rebellion, but the men who led these forces had been for the most part civil officials with the usual background and training. The major political actors now were primarily military men, the so-called war lords. They had risen through the new armies founded in the aftermath of the Sino-Japanese War, the most notable group here being the "Peiyang clique" of generals, early protégés of Yuan Shih-kai. From among this group came five future presidents or acting chief executives, a premier, and a majority of the war lords of north China.[86]

The first serious effort to reestablish political unity was carried out by Chiang Kai-shek. Originally a soldier, Chiang subsequently rose to the leadership of the conservative faction of the Kuomintang, the ruling political party, eventually taking over control of it by 1927. Although Chiang was able to subjugate most of the regional political and military power bases in the country, one dissident force led by Mao Tse-tung succeeded in escaping his determined efforts to crush it. Where the Kuomintang, after first championing a Westernized liberal ideology, tended under Chiang to fall back on a reassertion of many supposedly orthodox, Confucian principles, Mao purported to be making a clean break with the past. How much of his ultimate triumph is to be explained in terms of his ideological innovation and how much by his skill as a political and military tactician is a matter of some dispute. Whatever the answer here, the Communist Party under Mao was—in terms of its organizing principles and often its mode of action during the struggle for power—much more akin to an army than to a political party.

6 *The Armed Forces of Japan during the Meiji Restoration and After*

According to a widely accepted convention, the modern history of Japan began in 1853 when a small American naval squadron under the command of Commodore Matthew C. Perry arrived there. Dropping anchor in Edo Bay, Perry had come empowered to negotiate a diplomatic and commercial treaty with the government. The treaty would have the effect of obliging Japan to end her long-maintained policy of isolation from the rest of the world. Unwelcome as the incursion may have been, the Japanese did not have the means to prevent it. When the "black ships" of the American squadron, far more formidable than any vessels the Japanese had seen heretofore, steamed into Edo Bay against the wind, the size and range of their guns enabled them to disregard the defensive emplacements guarding the approaches to the chief city and political center of the country.

Although the Japanese leaders were not unaware of what was happening abroad, so graphic a demonstration of Western military power and technological virtuosity made a deep impression on them.[1] It also contributed to setting in motion a train of events which would lead within fifteen years to the downfall of the existing regime and its replacement by one established on very different political bases. In the estimation of those engineering this radical, even revolutionary transformation, the most pressing order of business was to repair the evident deficiencies in the defense posture of the country. They saw the task primarily as a matter of obtaining up-to-date weaponry of the kind possessed by the intruders from the West and then of training and organizing forces for its most effective utilization. Unlike many from so-called traditional societies aspiring to reorganize their military establishment along modern or European lines, the men of the new ruling elite in Japan were conscious practically from the start of the ramifications the reforms might have for the rest of society. Recognizing that the military power of the European states ultimately depended on the existence of a strong economy and an educated, vigorous populace, they initiated a fundamental reshaping of many of the institutions of Japanese society in accordance with European models. That they were able over a period of four decades to transform an apparently backward and feeble realm into one capable of competing on the world scene with the great powers is a measure of their accomplishment. It also constitutes one of the salient facts of the history of the nineteenth and twentieth centuries.

The political configuration of the country "opened" by Perry in 1853 had what might seem to Western eyes a number of anomalous features. Japan was, in theory, ruled as a single political entity by an emperor, the titulary of a line stretching back over two millennia, but some 800 years before, he had come to be little more than a figurehead. Real political power was wielded by the shogun, the chief military deputy of the emperor. For two-and-a-half centuries prior to the arrival of the Americans the shogunate had been vested in the Tokugawa house, rulers of the most extensive and powerful of the *han*, or feudal domains, into which the country was divided. Varying greatly in population, territorial extent, and wealth, each of the more than 250 domains was to all intents and purposes autonomous. It was to the feudal domain that an individual Japanese directed his loyalty and in terms of the domain that he defined himself politically. Originally the feudal domains had been organized above all for war, and endemic internal conflict had been a notable feature of Japanese life as far back as the thirteenth century. After an especially intense era of civil war lasting several generations, the Tokugawa house and its allies were able to subjugate or neutralize the other domains, a process climaxed by their victory in the great battle of Sekigihara in 1600.

As a consequence of its triumph, the Tokugawa house was able to bring about a far-reaching redistribution of territory among the domains, assuring itself a preeminent situation such that it exercised hegemony over the country. The Tokugawa house held direct control of some 25 percent of the land and with it the agricultural wealth of Japan, while domains allied to it comprised another 30 percent of the territory. As for the remaining domains, they tended to be located in peripheral areas and in general were either too distant, too weak, or too divided among themselves to be a significant threat to Tokugawa preponderance.

Having established their hegemony vis-à-vis the other domains, the

Tokugawa leaders were determined that it endure. The shogunate created a system of control to ensure that the balance of political forces within the country should never be to its disadvantage. The Tokugawa authorities took it upon themselves to oversee such matters as marriages among the *daimyo,* the heads of the feudal domains. *Daimyo* were required to spend every other year at the shogunal court in Edo, while their wives and immediate families had to reside there permanently, hostages to their good behavior. Any kind of official contact or relations between the individual domains were forbidden unless carried on through Edo. A pervasive and efficient secret police was set up to exercise surveillance over the behavior of the people lest elements of sedition take root. In general the shogunate sought, successfully for the most part, to maintain peace and order by depriving the various *han* of their discretionary ability to make war on each other.

Social stability was seen as a guarantee of order and the maintenance of internal peace. To promote that stability, the caste structure as it existed at the end of the sixteenth century was made permanent and given legal sanction. Movement between the castes was meant to be terminated. The warriors, or *samurai,* constituted the leading caste, followed by the peasants, the artisans, and finally the merchants. If in the domestic upheavals of the relatively recent past, the bearing of arms had been by no means the exclusive prerogative of the samurai, it became so under the Tokugawa shogunate, as peasants were deprived of the right even to own weapons.[2]

In the age of peace which had opened with the ascendency of the Tokugawa house, the samurai lost their primary purpose as fighting men. These now-superfluous men at arms might have become a socially disruptive element, had not a new function for them developed. It was from the ranks of the samurai that administrative personnel were recruited for both the shogunate and the individual domains. One noteworthy by-product of the process of turning the warriors into a caste of bureaucrats was their increasing literacy. Where in the early years of the Tokugawa shogunate a samurai who knew how to read and write had been a rarity, two centuries later most of them had acquired these skills, and "few would confess to the inability to turn a Confucian phrase."[3] Indeed, in a political system dedicated to the preservation of the status quo, Confucian learning and modes of administration, first introduced from China centuries before, had become most appropriate, and the official schools of the various domains sought to instill them in the new bureaucratic personnel. By the nineteenth century the process of transformation had so well succeeded that a large percentage of the samurai, while still taking an interest in military matters and retaining some outward attributes of a warrior caste, namely the wearing of their traditional two swords, had become quite civilianized.

Because ideas and artifacts from abroad were seen as disturbing to the status quo, contact with the outside world was meant to cease. Self-imposed seclusions became a major feature of Tokugawa policy, with no foreigner being allowed to enter and no Japanese to leave the country. To reinforce the ban on overseas travel the shogunate placed strict limits on the size of ships to be built. Over the succeeding centuries, the strictures of the shogunal authorities against any contact with foreigners came to be accepted as perfectly proper. One consequence was that shipwrecked sailors were now looked upon as dangerous trespassers and were apt to be treated in an unfriendly manner. That peace and tranquility could be so long maintained in what had been a notably violent society represents a major achievement. Originally a bellicose, adventurous people—the seafarers of East Asia—the Japanese through long years of repressive but undeniably effective Tokugawa rule evolved pervasive traits of decorum and an elaborate formality of behavior.

Formidable as had been the system of rule founded by the great Tokugawa shoguns, it was losing its effectiveness by the nineteenth century. The shogunate was in fact becoming dilapidated and increasingly unable to cope with, let alone master, the new social and economic forces that were emerging in the country. The decades of peace fostered by the Tokugawa regime had been conducive to the development of commerce on an ever-wider scale and with it the growth of a monetary economy, something favorable to the merchant caste but not to the traditional holders of power.

In Tokugawa Japan, political power was based on agricultural wealth, namely, the production of rice. The greater the size of a domain, the more rice its peasants could grow, with some 30 percent being taken by the domain authorities as tax revenues. It was in terms of a given quantity of rice that the stipend furnished to a samurai by the domain was calculated. Although the income of the samurai consisted of specific quantities of rice, many of their needs required that they have cash, and to obtain it they had to dispose of a portion of their stipendiary rice, usually at disadvantageous terms.

An apparently unavoidable corollary of the growth of a commercial economy was gradual but persistent inflation, which simply made worse the predicament of the samurai. Frugality was a traditional samurai virtue, but the practice of that virtue in the most dedicated fashion could not keep the former warriors from falling inexorably into debt and even into the direst poverty. Where the officially sanctioned caste structure had once been congruent with the realities of socioeconomic power, by the nineteenth century it was not so any longer. The workings of the economy advanced the interests of the caste lowest in the ascriptive hierarchy at the expense of the one at its head.

The economic problems besetting individual samurai tended to reflect in microcosm what was visited upon the domain governments. Dependent upon the product of a relatively static agricultural economy, they could only augment their revenues to meet rising expenses through a more merciless exploitation of the already overburdened peasantry, a process likely to provoke rebellion. The situation of the Tokugawa house was aggravated because the most intense commercial activity, with its inflationary ramifications, was to be found in its lands. In the peripheral domains, far from the centers of commerce, the pressures of inflation were less in evidence. The Tokugawa shogunate was thus by the middle of the nineteenth century in a more and more feeble condition. Beset by chronic financial difficulties, the shogunal government no longer displayed the administrative vigor it had once possessed, nor was it capable of exercising towards the other feudal domains its old, unquestioned authority. Shogunal preponderance now rested as much on no one's having yet found the occasion seriously to question it as on anything else.

Even at the height of their power, the shogunal authorities had never pretended that Japan's seclusion was complete and absolute. Nor indeed did they want to be ignorant of all that was going on outside of the country. From the earliest days they had permitted a small colony of Dutch traders to reside on an island in the harbor at Nagasaki, serving as agents for the transmission of information about the rest of the world, especially Europe. European sciences and technical studies came to be known as "Dutch learning," and a few scholars sought to master the Dutch language in order to partake of it. Appreciating the possible worth of some of the products of European intellect, especially for the defense of the country, the shogunate had in 1811 established an office for the study and translation of useful books from the West.[4] Even though it had endeavored, as a measure of internal pacification, to get rid of the modern, European-style weapons introduced by the Portuguese and others during the civil conflicts of the sixteenth century, the shogunate had still continued to purchase guns and mortars from the Dutch.[5]

The interest on the part of some scholars in European technical and military matters was further stimulated by the victory of the British over the Chinese in the Opium War. Experiments were initiated in several domains working towards the manufacture and the utilization of Western-style weapons. Some of these activities seem to have been encouraged by the Tokugawa authorities, while others were undertaken in studied disregard of shogunal disapproval. A few men were removed from office as a result, and some were even imprisoned.[6] It should be noted, however, that the shogunate made no serious, concerted effort to introduce Western weapons into the existing forces or to train men in their use.

The effect of the Perry visit on the rulers of Japan was one of profound dismay. Although the shogun was, in theory, the military chief of the country, bearing the title of "barbarian-subduing general," he could do nothing to prevent the Americans from landing and delivering a letter from the President of the United States. Having obliged the shogunal authorities to receive his embassy, Perry sailed away, promising to return the following year to negotiate a treaty establishing regular diplomatic and commercial relations. About all the shogunate was able to do against that ominious prospect was to attempt to improve the state of readiness of the coastal defenses by constructing new fortifications, and to lift the restrictions on the size of ships that could be built in Japan. Domains wishing to do so were also allowed to purchase warships and guns from Holland.[7] Indicative of the unprecedented gravity of the situation was the circular addressed by the shogun to all of the *daimyo* requesting them to submit their views on the reply to be made to the Americans. That the Tokugawa shogunate found it necessary to consult the heads of the feudal domains concerning the kind of matter over which it had heretofore exercised dictatorial authority was symptomatic of the erosion in its power and its self-confidence.

Despite the preference of a majority of the *daimyo* for maintaining the policy of seclusion, they were generally against any actions leading to overt hostilities with the Westerners. The shogunate thus had no choice but to sign a treaty opening commercial relations with the United States following the return of Perry in 1854. Four years later, in 1858, when the Americans along with the other Western powers pressed for the negotiation of a full-fledged commercial treaty, the shogunate sought to temporize and to rally support among the *han* for what it saw as an inevitable widening of relations with the foreigners. Once again it consulted the *daimyo* concerning the issues, and once again the replies were almost uniformly negative. Hoping to create a consensus in favor of the treaty, the shogunal authorities endeavored to win over to that point of view the imperial court ensconced at the traditional capital city of Kyoto. Much to their discomfiture, they found that the court favored continuation of the policy of seclusion, so they could only sign the treaty anyway and seek somehow to compel its grudging consent.

By the start of the nineteenth century, the emperor had long been a political cypher, all but forgotten by most Japanese. Still, the doctrine of imperial sovereignty had always been latent in Japanese thought, and a few had maintained interest in the imperial institution.[8] Research into the early history of the country, which had been carried on for several generations under the auspices of the Mito domain, stressed the antiquity and continuity of the imperial system. Of little immediate political importance in themselves, the efforts of the Mito historians could still remind those distressed at the decrepitude of the Tokugawa system that an alternative form of political organization was already in existence, at least in skeletal form. In that sense they made a significant intellectual contribution to the imperial revival slowly taking shape, and the shogunate simply added to the strength of that imperial revival when it sought to reinforce its shaky political situation by consulting the emperor on the subject of the commercial treaties.

Having signed the commercial treaties, the shogunate sought to console itself and mollify its opponents with the assumption that they were merely temporary and would eventually be abrogated. The foreigners would be expelled, the ports would be closed to external commerce, and Japan would return to the policy of seclusion considered by most to be proper.[9] To suggest that the shogunate work more vigorously to accomplish these desirable goals became a handy tool for many of its adversaries, especially at the imperial court, to use against it. Adding to the unpopularity of the treaties was their inclusion of the demeaning provision of extraterritoriality for all foreigners resident in Japan.

Almost immediately after the commercial treaties were signed, foreign merchants began to arrive in Japan, and their numbers would grow steadily. The quantity of trade was not great, but it did have a disruptive effect in some areas of Japanese life. This was in part because of the unfamiliarity of the Japanese with the trade practices and conventions of the West and their inability to oblige the foreigners to abide by the more restrictive commercial customs of the country. Thus the value of silver in relation to gold was set too high in

Japan, at least in comparison to the ratio accepted by the rest of the world. The result was a momentary but sharp outflow of the more precious metal during 1859. Then too, foreign demand for commodities like tea and silk made them increasingly expensive, aggravating the situation of the hard-pressed samurai trying to exist on their fixed stipends.

Already incensed at the humiliations they saw inflicted upon the country by the foreign powers, many of the samurai were quite ready to strike back at the people considered to be responsible for their financial distress. There were a number of physical assaults on Westerners and several were assassinated.[10] If actions of that nature were quite consonant with the traditional samurai code of behavior, the Western governments found them to be unacceptable. Assuming that the shogunate exercised authority over the whole country and was therefore responsible for the maintenance of internal order and the protection of foreigners resident there, the Western governments demanded it pay compensation for the attacks on their citizens. Already in serious financial difficulties, the shogunate did not welcome having to pay large indemnities for things done by people over whom it had no real control.

Whatever the economic or other forces weakening the shogunate, the chief manifestation of its debility was military. The Tokugawa authorities had made no serious efforts to repair the defense posture of the country even after the British victory in the Opium War suggested that Japan might well be the next target of Western imperialism. Nor did the Perry visit provoke the shogunal leaders to any vigorous measures. Part of the reason for this relative passivity may have been their consciousness of the disparity between the military force at the disposal of Japan and that of the Westerners. Some domains were in fact more eager than the shogunate to undertake military reforms and to experiment with Western military techniques. As a result, the clear military superiority over the other domains which the shogunate had possessed a century earlier was steadily eroding, until its armed forces were only slightly more imposing than those of its larger potential rivals. The authority exercised by the Tokugawa vis-à-vis those domains fell in proportion.[11] Not until the mid-1860s, a decade after the appearance in Japan of the Perry squadron, did the shogunate begin to evince a serious interest in the renovation of its military forces, and only in January 1867, with the arrival in Japan of a French military mission, were systematic efforts undertaken to organize and train an important part of the shogunal army in accordance with modern European military methods.[12] By that date, the balance of military and political forces had shifted radically to the disadvantage of the Tokugawa regime.

In the resistance movement that brought down the Tokugawa shogunate, the main initiative was taken by two domains, Choshu and Satsuma. A few of the other 260, notably Tosa and Hizen, contributed something to the antishogunal forces, but the great majority stood aside waiting to see who would be victorious in the struggle. There was no generalized, nationwide uprising. That the Tokugawa regime, despite its still-existing preponderance in territory and population, fell before the assault of so few of the feudal domains may be taken as indicative of how its morale and political power had degenerated by the middle of the nineteenth century. Still, decay within the shogunate should not cause the observer to underestimate the remarkable gifts displayed by its adversaries, especially the men from Choshu and Satsuma, or to disregard the special advantages possessed by these two domains.

Both *han* were comparatively big, with an unusually high percentage of their population being samurai. They thus had a relatively large pool of men available for either political or military action should the need arise. In contrast, the Tokugawa domain and its allies had fewer samurai than might have been expected, given their territorial extent. In both Satsuma and Choshu the samurai tended to be more traditionalist, more conservative, devoted to the old virtues of frugality and fidelity. These attributes were heightened by their being located some distance from the center of the realm. Both had for different reasons been able to achieve a reasonable measure of prosperity, contrary to the economic problems experienced by most domains in the mid-nineteenth century. This meant, among other things, that they could afford to purchase the Western arms which facilitated their triumph in the struggle against the shogunate.[13]

Satsuma and Choshu had long harbored antishogunal sentiments, having been on the losing side in the great struggle leading to the Tokugawa victory. These sentiments were particularly strong in Choshu, for the house of Mori, the hereditary rulers of the domain and until then the second greatest feudal power in Japan, had seen its lands reduced to one-fourth of their previous extent as a consequence. Anti-Tokugawa ceremonials and rituals formed an important element in the lore of the Choshu samurai.[14]

The history of Japan during the decade after the signature of the commercial treaties in 1858 was characterized by a complicated interaction between what might be considered national and local politics. On the national level the main contenders were the imperial court and the shogunate, each backed by those domains which were in its camp out of a sense of self-interest or for reasons of traditional loyalty. The ostensible issue here was whether the policy of seclusion should be reinstated, and with it the obligation of all foreigners to quit the country—a program of action reputedly championed by the court—or whether the opening of Japan should be accepted as inevitable and relations with the outside world extended—the policy advocated by the shogunate. The real point at issue, however, concerned the distribution of power between the two institutions. Would there be a restoration of the shogunate to something like its former position of hegemony, or would the imperial revival already under way before 1853 be carried even further, at the expense of the power and authority of the Tokugawa regime?

Political developments on the national level had their reflection in the internal politics of various domains. Advocates of the imperial cause, the shogunal cause, or something in between vied for control over *han* policies in such a way that national questions were coming to supersede purely local matters as the primary determinates of political action, at least in a number of the domains. Further, not only did the great issues of the proper policy for the nation to follow have reverberations on the local level, in that they increasingly determined who would rise and exercise leadership within a given domain, but something like the obverse also tended to be true; that is, the results of the factional struggles within certain domains had a major influence on how the balance of forces worked out on the national level. Some of the more ambitious or adventurous of the samurai from Choshu and Satsuma saw the turmoil over the commercial treaties as furnishing an opportunity to advance the political fortunes of their particular domain. They would seek to promote a course of action which could be interpreted as redounding to the good reputation and influence of the domain on the national level and which would add to the discomfiture of the shogunate, something they found congenial given their traditional anti-Tokugawa sentiments.

Choshu was the first to come forward, dispatching representatives to the imperial court to advocate mediation between Kyoto and Edo. This policy was denoted by the slogan of "the Union of the Court and the Shogunate." Implied here was the desirability, indeed the necessity, of restoring to the emperor some of the power and authority reputedly usurped over the centuries of the shogun. The champions of that cause came to be known as imperial loyalists. Choshu did not long remain alone in its efforts, for men from Satsuma arrived at Kyoto in the summer of 1862 with the same avowed intention. During the years 1862–1863 representatives from the two domains engaged in a complex political competition whereby each tried to arrange the acceptance of its own version of the union of the court and the shogunate.

The rivalry between Satsuma and Choshu for predominant political influence at Kyoto led each to call for the enactment of ever more daring and extreme programs in the name of the emperor. In the process, the representatives from Choshu went the farthest, ultimately advocating a policy that could be summarized in terms of the slogan "Honor the Emperor and expel the barbarians." This was something far more radical than had been the intention of the Chushu leaders when they had originally sanctioned efforts to promote a policy of mediation between the emperor and the shogun.

Since the signing of the commercial treaties in 1858, the court had ostensibly regarded them merely as temporary expedients and had prodded the shogunal officials as to the desirability of terminating them. A corollary of such an act was a return to the policy of seclusion and the ejection of the foreigners now present in Japan. Under the influence of Choshu loyalists the court was finally moved to order the shogun to set a date upon which the foreigners would begin to be expelled. So great had been the decline in the vigor and political authority of the shogunate that it could only acquiesce, whatever its misgivings about the possibility of implementing the policy. The shogunate therefore announced a specific date for the expulsion and informed the other domains of the desires of the court. Alone of all the domains, Choshu prepared to act in accordance with the stated will of the emperor. It is impossible to know if Choshu authorities really believed the imperial order could be carried out. Perhaps the fervor of the young Choshu loyalists was so intense that they were able to ignore political and above all military reality.

On June 25, 1863, the day set by the shogunate for the expulsion of the foreigners, Choshu batteries opened fire on an American ship passing through the Straits of Shimonoseki. There being no foreigners in Choshu to oust, the most that could be done was to attempt to keep foreign ships from coming too close to the shores of Japan. A French and a Dutch vessel were subjected to the same treatment during subsequent weeks. By way of reply to the provocative actions on the part of the Japanese, in the middle of July an American warship shelled the forts overlooking the straits and sank two gunboats recently purchased from abroad by Choshu. A few days later a French warship fired on the forts, after which a contingent of troops went ashore and proceeded to destroy them.[15] Even after that military misfortune, Choshu batteries continued over the ensuing months to fire intermittently at foreign vessels passing through the straits.

The Western powers were not pleased to have ships of theirs which were engaged in peaceful commerce fired upon by the Japanese. They held the shogunate responsible for the behavior of a subordinate domain and sought to have it stop the hostile actions of Choshu. Following the evident inability of the shogunal authorities to comply with their wishes, the Western powers decided to take action on their own initiative.[16] On August 15, 1864, a fleet of seventeen ships from Britain, France, Holland, and the United States opened fire on the Choshu forts. These were effectively destroyed once again while a landing party decisively defeated Choshu troops in open battle. The obstinate domain was subsequently obliged to agree to leave the straits unfortified.

At about the same time that Choshu first chose to engage the Western powers in hostilities, with such adverse results, Satsuma also received a lesson concerning the efficacy of European methods of waging war. On August 15, 1863, seven British warships bombarded the *han* capital of Kagoshima, leveling a large part of the city, after they had been fired upon by shore batteries. The British had come to enforce a demand for monetary compensation from Satsuma after samurai from that domain had killed a British subject the year before.[17] The leadership of both Choshu and Satsuma had to recognize how little they might hope to accomplish militarily against the West given the present state of their armed forces. The reaction of Choshu to that demonstration was particularly noteworthy.

Like other domains in mid-nineteenth-century Japan, Choshu had devoted some attention to the study of "Western" or "Dutch" learning, although to a lesser degree than several of them. Following the visit of the American squadron under Perry, the interest in these questions had become more intense with the introduction into the curriculum of the *han* school of geography, astronomy, history, and military science, but the number of those interested in the new learning was very small compared to the people devoted to swordsmanship and the study of Confucian classics.[18]

No serious attempt was made to organize more than a fraction of the military forces of the *han* along modern lines until the armed confrontation with the Western powers. At that point, the domain authorities recognized serious military reform would have to be undertaken if Choshu were to defend itself against the possible encroachments of the foreigners. Accordingly efforts were undertaken to organize a new military force. It was meant to be a kind of supplement to the regular units of the domain's army, but rather than relying exclusively on the members of the warrior caste, the new force would draw its members from all social strata, commoners as well as samurai. The only criterion for the men being recruited was that they possess courage and resolution. These new soldiers, numbering about three hundred, would be trained in the use of both Western and traditional Japanese weapons, but with the emphasis on the former. They were called the *Kiheitai,* meaning "strange" or "surprise" troops, indicating that they were different and set apart from the regular *han* forces. The *Kiheitai* served as the forerunner or model for other auxiliary military units, known as *shotai,* not less than 155 of which had been organized within a couple of years.[19]

The Tokugawa authorities could hardly be uninformed concerning the military developments in Choshu or pleased at the idea of recruiting commoners for military service. The shogunate was also angered at the extremist policies advocated by the Choshu representatives at Kyoto and by the unruly, even violent, behavior of some of its samurai at the court. In the estimation of the shogunate the time had come to teach a lesson to the insubordinate domain and to demonstrate the penalty attached to its outrageous behavior. To that end, the shogunate organized an army of some 150,000 men drawn from the Tokugawa lands and from vassal and collateral domains. Faced by so formidable an array Choshu had no choice but to back down. By the peace terms it accepted in January 1865, the Choshu government had to admit the error of its ways and to dissolve the new military units, after which the shogunal forces disbanded. But the leaders of the *shotai* refused to accept dissolution, resisting by force the directives of the more conservative men in the leadership of the domain. There followed what has come to be known as the Choshu Civil War. In this conflict, the new military units defeated the regular, traditionally organized forces of the domain. As a result, the imperial loyalists were able to displace the conservatives in the political leadership of Choshu.

Given the incorrigible behavior of the domain, the Tokugawa government believed that it had no alternative but to try again, only now with the intention of inflicting a more serious penalty, possibly even the destruction of Choshu as a political entity. The prospect of a second attack by the shogunate within two years sparked the Choshu government to undertake a complete reshaping of its military forces. Responsibility for overseeing the task was given to Omura Masujiro, a long-time student of Western military practice and a person endowed with notable talents as an organizer. Where the new-style military units had originally been looked upon as auxiliaries of the regular Choshu army, they now became its core. The threat of the subjugation and even the possible destruction of the domain added a sense of urgency to the military preparations undertaken by Omura and the other loyalist samurai. The new formations were reorganized as standard units of 150 men, armed with rifles.[20] All classes, be they privileged or commoner, once mobilized, were formed into rifle companies and drilled as disciplined, tactical units.

The reorganized armed forces of Choshu represented something unprecedented in Tokugawa Japan. They went counter to the principles of the samurai, not to mention the self-image of those in the warrior caste. Trained to fight with the sword, and disdaining firearms as weapons for men of lower status, the samurai were proud of their archaic skills and anxious to display them in single combat. Now, however, the samurai carried rifles on the same footing with commoners, alongside of whom they fought in homogeneously organized groups at the orders of a single command authority. As for peasants and other commoners entering the new rifle units, they were pleased to obtain the status of warriors and to enjoy some of the attached perquisites, a surname and the right to wear a sword.[21] The military transformation of Choshu in accordance with European standards turned out to be a development of far-reaching significance for Japanese society, but the men who launched the reform proper did not do so with any radical social goals. They were simply trying to enlarge and strengthen the armed forces of the domain in order to defend it against its numerically superior adversaries.[22]

When in August 1866 the shogunate launched its second punitive expedition, the forces it mobilized were far larger than those of Choshu, but with regard to troop morale, weaponry, and organization, they were deficient. Within a few months the shogunal host was driven back on all fronts. The sudden death of the shogun allowed his forces to sign a truce. On the field of battle, then, one domain had been able to defeat the shogunate. Choshu was much aided in its struggles with the shogunate by the benevolent neutrality of Satsuma, a policy championed in particular by its recently elevated, charismatic military leader, Saigo Takimori.[23]

The Choshu victory was followed by an interim period in which the major political elements in the country, the shogunate included, tried to evolve a new system for managing their mutual relations. What impressed many observers was the apparent rapid revival of Tokugawa vigor and initiative under the new shogun, Keiki. Supported politically and financially by France, and under the guidance of the minister from that country, Keiki initiated a major reorganization of the shogunal system of administration and of the armed forces in accordance with what were theoretically French practices.[24]

Faced by this threatened restoration of shogunal power, Choshu loyalists, now with the active support of allies from Satsuma and a few other domains, moved rapidly and decisively to protect what had been gained by force of arms. They sent troops to seize the palace at Kyoto. Then, armed with an imperial rescript or decree obtained under somewhat irregular circumstances, they announced that the Emperor had reassumed the powers and prerogatives he had once possessed and that the shogunate as it had existed over the past two-and-a-half centuries was at an end. Tokugawa forces resisted the coup, but to no avail. In two battles fought near Kyoto they were defeated, even though they were in good spirits and outnumbered the loyalists by a considerable margin.[25] Following their victory, the loyalist troops marched toward Edo. To spare the country the depredations of an extended civil conflict Keiki resigned as shogun, while also accepting a drastic reduction in the size of the Tokugawa holdings. The Tokugawa house was thus deprived of its overwhelming territorial preponderance as well as the potential resources whereby it might hope to reassert itself militarily and politically in the future should the opportunity arise. Some supporters of the Tokugawa cause did not willingly acquiesce in the settlement accepted by Keiki. There was sporadic but bitter fighting in the northeastern part of Honshu for much of the rest of the year before they were finally defeated, while resistance on the part of a few die-hards on the island of Hokkaido did not end until the following year.[26]

Much of the success of the anti-Tokugawa movement may be attributable to the high morale of its adherents, which was based on their belief in traditional values and their devotion to the imperial institutions embodying them. But to transmute those attributes into real political power required the use of armed force, and in the case of Choshu that was furnished by the *shotai,* military units organized and trained according to European principles. Samurai from Choshu and Satsuma had had first-hand experience concerning the effectiveness of the Europeans' weaponry and military techniques. They were more than willing to learn from that experience, to accept the empirical lesson administered and, in the hope of defending the country, "to use the barbarian to control the barbarian."[27] Special political circumstances dictated that the first use of the European-style forces they had organized be not in defense of Japan against the foreigners but in a civil conflict against other Japanese.

The situation of the new leaders of Japan, the authors of the transfer of power which has come to be known as the Meiji Restoration, was quite insecure following the overthrow of the shogunate. There were no effective instruments of power for them to take over in the aftermath of the collapse of the shogunate, since whatever organs of rule or methods of central administration may have existed under the Tokugawa regime had disappeared with its collapse. In June 1868 a council of state was set up as the supreme organ of government for the country, but its powers were fluid and ill defined. Aside from the prestige its members may have possessed because they had been on the winning side in the recent civil war, it had little legitimacy or real authority over the country. Above all, the new government possessed no regularly established sources of revenue and no military forces of its own. The troops who had accomplished the overthrow of the shogunate and brought the loyalists to power were under the command of their individual domains. Most had been marched home once the fighting was over.

The major problem immediately confronting the new government stemmed from the particularism inherent in the traditional feudal order. The Meiji leaders recognized that a more effective defense of Japan against foreign adversaries, the ostensible reason for their having risen against the Tokugawa regime, presupposed greater political unity within the country and thus a concomitant weakening of the feudal domains. It was not a course of action welcome to most of the imperial loyalists. Indeed, the successful struggle against the shogunate had led to a resurgence in *han* spirit and pride, especially among those from the victorious and now-dominant domains like Choshu and Satsuma.

The ardent young loyalists were, after all, samurai who had grown up under the feudal system and had been educated according to its values. Devoted as they may have been to the cause of the emperor, they were also faithful servitors of their respective domains, and they saw no incompatibility between the two ideals. They had certainly not taken part in the struggle against the Tokugawa regime with the goal of destroying the whole existing structure of social power and political authority, one to which they were psychologically committed. Rather, the imperial loyalists had aimed at terminating the hegemony incompetently exercised by the Tokugawa house and replacing it by a political system capable of better mobilizing the energies of the country to resist the encroachments from abroad. They wanted no greater changes in the sociopolitical structure than would be sufficient to promote a militarily strong and economically healthy country, "to prosper the state and strengthen the armed forces." They were to find, however, that any measures leading to a real increase in the military capabilities of the country would result in the emergence of a very different Japan.[28]

The most expeditious way to provide the imperial regime with an adequate military force would be to persuade those domains having the largest and most obvious stake in the Restoration to assign the government a contingent of their troops. This presupposed that the people involved trusted each other and were in agreement. Unfortunately friction had developed soon after the Restoration between Saigo, head of the Satsuma forces, and the other members of the government, with the result that he ceased to participate in its deliberations. So long as Saigo, now the leading military figure in the country, boycotted the government, there was little chance that Satsuma, the domain with the largest contingent of soldiers, would cooperate in the plans for an army under the control of the government. The other Meiji leaders believed that the participation of Satsuma was absolutely necessary and that Saigo had to be involved, if for no other reason than the personal authority he would impart to the undertaking.

The issue as it was presented to Saigo was stark and simple. If the domains were to contribute contingents for the formation of some kind of centrally controlled armed force, they would no longer be under the command of the domain authorities. As it was stated to him by one of the leading young loyalists: "Satsuma men must be willing to turn against the Satsuma lord if the occasion demands it."[29] Saigo was finally won over, and the way was cleared for the creation of a sizeable military force under the exclusive control of the government. On April 17, 1871, it was announced that an imperial body guard, *goshimpei,* of some ten thousand men had come into existence, consisting primarily of contingents from Choshu, Satsuma, and Tosa. It was a big enough force for the government now to be able to assert itself, at least against any foreseeable opposition elements within the country. That the next major move undertaken by the new regime was the abolition of the feudal domains would not seem to have been fortuitous.

To eliminate the *han,* or even to reduce their power in any marked fashion, was a forbidding prospect, if only because they had provided for centuries the framework for the public life of the Japanese people. Nevertheless, almost from the moment of the Meiji Restoration, a few hardy and prescient men set about trying to persuade important personages in their domains of the desirability and even the absolute necessity of effective political unity for the country. Only a small minority of the samurai were ever won over to a belief that the feudal order in Japan had to be terminated, but that minority contained most of the leaders of the class.[30]

The persuasive efforts of the advocates of political unity finally, in March 1869, led the *daimyo* of Choshu, Satsuma, Tosa, and Hizen to offer their domains to the emperor, an action known as the "return of the *han* registers." A number of other *daimyo* quickly followed suit, possibly not wishing to be left behind in any distribution of rewards. Once it had accepted the lands thus offered, the government invited the heads of the remaining domains to do the same thing, promptly appointing them as governors of their domains in recompense.[31] If their power over their lands did not appreciably change, they nevertheless now held office at the behest of the central government rather than as a consquence of the play of local political forces. While concerted efforts were under way to convince people of the necessity of ending the system of decentralized political power, the imperial government was increasing its pretensions vis-à-vis the domains through a series of singly unimposing but cumulatively significant measures. Thus the government had already gone a considerable distance toward ending the autonomy of the domains when, in August 1871, it announced their dissolution.

The abolition of the domains was met with a sense of shock on the part of the samurai, but aside from a few isolated acts of terrorism, they did little. What some have called the Second Restoration was carried through with ease, despite the apprehensions of the imperial government about the potential repercussions of its bold policy. The very fact of the domains' autonomy, and the antipathy that many of them had always felt toward each other, interfered with their now acting in concert and raising significant resistance to the actions of the imperial government. Of course, the generous financial settlement made by the government certainly helped to pacify their leaders. Most of the domains were on the verge of bankruptcy by the end of the Tokugawa era, and the willingness of the government to assume responsibility for paying many of their debts was most welcome.[32]

The *daimyo* did particularly well in this political transaction. They were to receive as an annual payment the equivalent of 10 percent of the gross revenues of their domains while at the same time being relieved of managing the financial burdens of governing them—in particular the paying of samurai stipends. As for the rank and file samurai, they were treated with less consideration. Their stipends did continue to be paid, but at a reduced rate. In the final analysis, the chief reason why the abolition of the domains was achieved with so little incident was that the central government possessed sufficient force to be certain of getting its way, even if the imperial bodyguard only constituted an army in embryo. For the country really to be defended, that embryonic army would have to be more fully developed.

The armed struggles leading up to the Meiji Restoration had amply demonstrated the superiority in combat both of Western weapons and of the tactical formations created to exploit them. That would seem to provide sufficient reason for the armed forces of the imperial regime to be organized along European or Western lines, but how to provide the personnel for this military body raised a number of political questions. There was in fact a large manpower pool already at hand—the four-hundred-thousand samurai still drawing stipends from the government. If the chief concern of the new government was to defend the homeland against the imminent threat posed by the foreign powers, a volunteer professional force raised from the traditional military caste might have been the most expedient means. Already imbued with an appreciation for martial matters, they would have required little training and indoctrination. Then, too, to have recruited the new armed forces from among the samurai might have helped to integrate that numerous and important social group into the imperial regime with as little disruption as possible.

There were, however, a number of reasons, both military and sociopolitical, for not relying too much on the samurai in manning the new army. Courageous as they might be, in their approach to war they were individualists, seeking to outdo each other in feats of personal valor in an age when circumstances placed a premium on the disciplined consistency of the fighting units in battle. Moreover, much in the political outlook and behavior of the samurai was likely to be determined by their having been brought up under the feudal system. Having originally sworn loyalty to the leader of a given domain, they could be powerfully predisposed in favor of people from that domain, even if it no longer had an official existence, to the detriment of their obligations to the imperial regime they were meant to serve.[33] To have turned to the samurai in recruiting the personnel of the new armed forces could well have led to the introduction of particularistic, even subversive, elements into a body upon whose loyalty, cohesion, and unity the new masters of the country had to depend. To allow the samurai to continue to fulfill their military function within society, more or less to the exclusion of other classes, would be somehow to perpetuate the inferior status of the latter at a time when the Meiji leaders were coming to believe that the unity of the country as well as its military strength would be better served by placing all members of society on an equal footing.

To a number of those in the ruling elite the only viable alternative to an army made up of samurai was one recruited through some form of conscription. Among the early champions of conscription, the most notable and influential was Omura, the first person to be head of the Military Affairs department in the new government. He advocated the introduction into Japan of something like the system of military service to be found in several European countries. His forthright espousal of his views on this subject may well have led to his being assassinated by some traditionalist Choshu samurai late in 1869.[34] The eventual successor to Omura as head of the Military Affairs department was Yamagata Aritomo, by birth a low-ranking samurai from Choshu.

More than anyone else, Yamagata may be credited with laying the foundations for an effective military force under the control of the central government. His evident military talents and his abilities as an organizer had been revealed during the struggles against the shogunate, when he had first come to the fore as leader of the *Kiheitai* and later as commander of new Choshu military formations in the campaign against the shogunate. That experience had provided him with ample evidence concerning the soldierly qualities of the peasants incorporated into the fighting units even if only on a temporary basis. It was obvious to him that the samurai did not possess a monopoly of the warrior virtues.

Yamagata was determined that Japan have armed forces capable of meeting Western standards of performance, and like Omura he was convinced that they should be founded on conscription. His convictions on the subject of conscription were reinforced by what he observed on the world tour he made in 1869–1870 at the behest of the government to study foreign military practices. Yamagata was especially impressed by the patriotic, militaristic spirit he sensed in Prussia. Universal military service contributed greatly to that spirit, he believed, by instilling the youth with soldierly habits and by making them active defenders of their country, and he foresaw conscription having an analogous effect in Japan.[35] The members of the new ruling elite all favored universal conscription, apparently even the traditionalist Saigo.[36] Thus an imperial edict calling for the introduction of conscription was promulgated in December 1872, to be followed by a law of January 10, 1873, providing that those inducted would serve three years of active duty in the army and four years in the reserves.

The proclamation prefacing the law purported to show that by instituting conscription the government was not copying European models. Rather it was returning to traditional Japanese practice whereby every peasant had once been a soldier when the need arose, hastening to serve his emperor. The permanent segregation of the warrior from other social and professional castes and their monopoly of the military function in society could thus be understood as a relatively recent and implicitly unhealthy innovation introduced by the Tokugawa regime. Now, as was stated in the words of the proclamation, both the samurai and other groups were to be understood "as people of the same imperial country and in the service of their country . . . there should be no difference between them."[37] Despite the supposed universality of the military obligation assumed by the young manhood of Japan, service in the armed forces at first was undergone by only a restricted portion of the populace. For a number of years the government lacked the financial resources to do more.

The new military system was not greeted with enthusiasm by the Japanese people. For a hard-pressed peasant household, already subject to heavy taxation, the prospect of losing the productive labor of a family member for a period of three years was an onerous burden.[38] Peasant resistance to the new military obligation was manifested in a number of ways. There was widespread rioting following the promulgation of the law, and it has been calculated that some fifteen uprisings in the early years of the Meiji era were caused by opposition to conscription. Some of those liable to military service went so far as to seek a physical exemption by mutilating themselves, while others simply fled their place of residence. As late as 1889, the number absconding annually amounted to about 10 percent of all the able-bodied males eligible for military service. Eventually the government recognized that the fervor of the opposition to conscription might be based on the people's anger at the more inequitable exemptions written into the system, and it set about eliminating them.[39]

More worrisome to the government than peasant discontent over conscription was the antagonism of the samurai. Where the peasants resisted the obligation to bear arms, the samurai objected to anybody but themselves doing it. For the samurai, the initiation of universal military service was symbolic of what they took to be their ill treatment at the hands of the new government. Many of them had committed themselves to the arduous struggle against the shogunate only to discover that their reward was to begin to lose their special marks of status and their time-hallowed privileges. In August 1871 the government called upon the samurai to give up their distinctive coiffure and style of dress. Then it suggested that they stop carrying their traditional two swords, finally forbidding the practice outright in 1876. By that time the situation of the samurai as stipendiary dependents of the state was also about to be terminated.

Continuation of the samurai stipends had been one means by which the government had bought off the warrior caste at the time of the imperial restoration and later when the domains were abolished. Because the stipends amounted to too great a burden for the financially fragile new regime, it commuted the payments in rice into government bonds on a voluntary basis in 1873. The commutation was made compulsory in 1876, although the amount of money received by the individual samurai in the transaction was not sufficient for any kind of decent standard of living. The more resilient of the samurai were able to come to terms with the policies launched by the government and to cope with the new conditions of life, but for many it was extremely difficult. The recently founded army and police force were not large enough for more than a fraction of them to obtain employment in areas they found most congenial. Unable to pursue careers compatible with their prior training or their particular outlook, they could only stand aside in frustration and anger.[40]

Samurai discontent over the policies of the new imperial regime posed a constant problem during the first decade of its existence. There were several outbreaks of rebellion on the part of disgruntled samurai which, for the most part, were suppressed with relative ease. The major exception was the 1877 uprising in Satsuma under the nominal leadership of Saigo. That was a far more serious affair. Alienated by the policies of the imperial regime toward the samurai, he had resigned his post as head of the armed forces and returned to Satsuma, where he had opened a number of schools for the teaching and preservation of the samurai way. So great were his magnetism and his reputation that by 1877 Saigo had attracted some twenty- to forty-thousand adherents.[41] To all intents and purposes the area comprising the former domain of Satsuma was under the control of Saigo and outside the jurisdiction of the Meiji state. An effort made by the government in the name of prudence to remove ammunition and weapons from the Satsuma capital of Kagoshima provoked the followers of Saigo to rebel. At that point he believed he had no choice but to place himself at the head of the uprising, even though he was probably convinced of its ultimate futility. He was to perish in the final battle of the rebellion.

Unlike previous outbreaks against the government, the Satsuma rebellion did not involve a mere handful of malcontents. The imperial regime was obliged to undertake a major campaign to suppress it, mobilizing some forty-thousand men, including the whole of the standing army as it then existed, whatever reserves were available, and a portion of the police force. It took some six months of hard campaigning and cost fifteen-thousand casualties, including six-thousand deaths, before the Satsuma defenders of the samurai way were overcome.[42] Armed chiefly with swords and antique matchlocks, the Satsuma rebels fought with notable valor, but intrepid as the samurai were, the decisive fact was that the imperial government was able to organize and equip forces of sufficient strength to overcome them. Any one of the soldiers mobilized by the government may have lacked the military skill and courage of an individual samurai, but the recently conscripted peasants showed themselves to be capable of becoming steady, solid troops who could support the rigors of a long campaign, fight in disciplined tactical units, and suffer severe casualties without breaking. The peasant soldiers more than vindicated the faith in their potentialities held by Yamagata and by Omura before him. The Satsuma rebellion was the last serious effort in the nineteenth century on the part of various groups in Japanese society to use violence to unseat the new regime or to make it change its policies.

For its first few years the imperial regime was in a precarious financial situation. The money to pay for ambitious undertakings like the organization of the armed forces and the assumption of the domains' financial obligations had to be raised through some questionable expedients. There were no regular sources of state income already in existence which could be tapped, nor were there any centralized financial institutions for the Meiji government to take over. During the feudal era just ended, public finance had been managed on a local, decentralized basis, with the authorities in each domain responsible for raising their own revenues from within their area of jurisdiction. It was not a system adequate to meet the growing demands of the imperial state. The government was thus obliged to develop a new fiscal system *ab ovo* and in some haste.

The basic source of state revenue was to be a land tax. Unlike previous taxes which had been levied in kind on the produce of the soil and which had varied in accordance with whether the harvest was good or bad, the new tax was monetary and calculated according to the assessed value of the land. It was not likely for that reason to undergo any great variation from year to year, thus allowing much greater predictability with regard to the revenues of the state, and the introduction by the government of rational budgetary practice. The new tax system also led toward more uniformity in the Japanese system of landholding. Because the tax was to be paid by the landowner, the principle of private ownership had to be established throughout the country, where in many areas such matters had hitherto been rather nebulous. In 1872 the government began to issue deeds to those having what seemed to be the strongest claims to title.[43] The land tax was to be the basic source of revenue for the state during the first several decades of the Meiji era, providing some 90 percent of the public money even in the 1890s after one generation of imperial rule. Twenty years later it still accounted for 60 percent.[44]

The fiscal demands made by the new regime on the Japanese people, the peasantry above all, were undeniably large. A portion of their income comparable to what had been taken by the domains in Tokugawa times now went to the imperial state, and the peasants were not likely to find the benefits to be derived from the new system of government commensurate with its costs. In this perception, they were much like the peasants in every society upon which modern public institutions were imposed. The peasants always seem to be the ones who bear a disproportionate share of the burdens of the process. Whatever objections the Japanese peasantry may have harbored were to no real avail. At most they could rise in rebellion—sporadic, spasmodic efforts doomed to almost certain failure, for the government now possessed in the new army a coercive instrument capable of asserting and maintaining its control over the people. In other words, the peasants paid a sizable share of their income in taxes to the government, much of it going to maintain a large military establishment, one major aim of which was to keep them docile and obedient.

In terms of its effect on the structure and functioning of Japanese society, the Meiji Restoration amounted to a revolution, and the single most revolutionary action of the new regime was probably the introduction of conscription. By eliminating the most significant of the privileges of the samurai caste—the exclusive right to bear arms—the state, through conscription, clearly furthered the equalization of legal and social status within the country. Conscription also contributed to a fundamental reshaping of the outlook of the Japanese people. A vital element in the strength of the nineteenth-century national state was a high degree of emotional unity among the great mass of its inhabitants, as well as their conscious commitment to the goals of the state as these were defined by the ruling elite. Attributes of this nature could hardly exist in Japan before 1868 since the country was divided into some 250 separate, all-but-independent polities—each with its own traditions and its own compelling corpus of loyalties—yet these attitudes had certainly come into being within a generation of the Meiji Restoration, and the creation of a mass army recruited through conscription and a nationwide system of education was essential to the process.

Military service was meant to be the climax to a process of indoctrination which started in schools. Under the educational system first established in 1872, all Japanese children had to undergo four (later six) years of primary schooling. In addition to acquiring at least the rudiments of literacy, they were also introduced to the official ideology of the Meiji state with its emphasis on loyalty to the emperor and to the nation. Education was to be considered less as a way to open the minds of the young students and help them to develop their individual potentialities than as a means of training technically competent members of the reorganized, modern society and instilling in them the correct, orthodox view. Carried further in the army, the whole process appears to have been very successful. Within a generation the average Japanese was imbued with a fanatical sense of national pride, taught to "glory in Japan's military tradition . . . [and] to believe that death on the battlefield for the Emperor was the most glorious fate of man."[45] Every soldier was taught that devotion to the emperor and obedience to his commands were imperatives that went beyond traditional obligations to family, clan, and village.

Along with what was accomplished through the government's conscious efforts at indoctrination, the very experience of military life itself had a profound effect on Japanese youth. A young man from the peasantry, rooted in a localistic, rural way of life, generally had to move to a city where the army barracks were located and to adjust to an urban environment. Unlike the closely knit existence of his family and village with their personalized obligations, life in the army and in the city was impersonal in its demands. To undergo it for the first time must have come as a shock for the young men from the countryside.

Once enrolled in the army, a peasant youth was exposed to a variety of new objects, practices, and ideas. He was housed in barracks equipped with beds, stoves, and eventually electric lights. Although his army pay was little more than a pittance, he had probably never before received a monthly salary.[46] Military uniforms were the earliest articles of Western-style dress to be adopted on a large scale in Japan, and it was through conscription that numbers of people were exposed to that mode of apparel for the first time. The principles of clothing design manifested in military uniforms were then introduced into work and school garments, thus becoming an everyday feature of Japanese life.[47] Military service was also likely to have an appreciable effect on the dietary habits of the Japanese, introducing people to the eating of rice, fish, and meat on a regular basis. A poll taken in 1892 revealed that some 70 percent of those inducted into the army claimed that army food was better than what they usually received at home.[48] Service in the armed forces clearly worked to expand the horizons of a young conscript. While under the colors, he met recruits from other areas in the country, and was thus made aware of the larger world beyond the confines of his native village.[49]

The new Europeanized army had a great influence on the economic development of the country during the early decades of the Meiji era. Determined to create an effective military machine as rapidly as possible, the ruling elite recognized that this necessitated the support of a "strong, well-integrated military industry."[50] Factories established to meet the material requirements of the armed forces formed the leading edge in the industrialization of the country. So marked was the importance of the needs of the armed forces that one authority has commented that "nearly all the mechanical industries which apply modern science and arts found their origin in military industry or developed under its influence."[51]

That the needs of the armed forces constituted a major factor in the origins and growth of heavy industry is something Japan shares with a number of other countries in the modern world. What is striking about the Japanese case is the way in which the growth in production of consumer goods, especially textiles, was also significantly affected by the demands of the armed forces. The traditional Japanese market for woven goods did not encourage mass production since the demand was for a great variety of fabrics, most of them capable of being conveniently manufactured through artisanal methods. The armed forces, however, required the production of cloth of a standardized kind in increasingly large quantities. A number of textile firms took advantage of that fact to the degree that much of the development of the modern textile industry, it has been argued, came about in order to meet military demand. An inevitable concomitant of the growth of the cotton textile industry and of its ability to produce cloth in vast quantities at a low price was the decline of small-scale production in the home.[52]

A number of factors led the government to assume a forward, active role in the industrialization of Japan. For one thing, the level of capital accumulation in the country was very low.[53] The Japanese people, either as individuals or in groups, were unable or unwilling to marshal the financial resources necessary for investing in industry. A thriving commercial class may have come into existence by the end of the Tokugawa era, but the merchants lacked a vigorous spirit of capitalistic enterprise and the willingness to assume the risks attendant on founding large-scale manufacturing firms. Only the government appeared ready to assemble the resources for the task. Then, too, the issue was one of national security, and it was to be expected that the government take the lead. The formation of investment capital in suitable quantities could not have been undertaken without the fiscal and financial reorganization of the country carried out in its early years by the new regime. Taxes provided most of the needed funds, and as noted above, the primary fiscal instrument was the land tax paid above all by peasantry.

Having supplied the initial impetus for the industrialization of the country, the state was prepared by the 1880s to turn over part of the plant it had created to a few trusted financial oligarchs, generally on most attractive terms. Nevertheless, it did keep control over certain strategic enterprises, such as arsenals, shipyards, and a few branches of the mining industry.[54] Despite the willingness of the state to divest itself of many manufacturing facilities, its preponderance in heavy industry persisted past the turn of the century. As late as 1907, government-run factories in that sector were employing some 60 percent of the workers and utilizing 70 percent of all horsepower available to industry, although the proportion devoted to armament production was declining as Japan began to engage in more varied forms of manufacturing.[55]

In their efforts to set up a military establishment capable of meeting contemporary Western standards of performance, the leaders of the Meiji regime turned to the West for help and guidance. During the mid-nineteenth century, France was considered to be the leader in matters pertaining to land warfare, so the shogunate sought the assistance of that country when in the 1860s it set about trying to improve the state of its armed forces. The government of Napoleon III, eager to extend its influence to the more distant corners of the world, had been glad to accede to Japanese requests, but by the time a French military mission had arrived and set to work in 1867, it was probably too late to prevent the military defeat and political collapse of the Tokugawa regime. Following the overthrow of the shogunate, the new Meiji government continued to make use of the services of France, even after 1870, despite that nation's military reputation having been somewhat tarnished in the Franco-Prussian War.[56]

Down through the 1870s the French mission worked at educating the Japanese in the fundamentals of modern military practice, and its accomplishments were considerable. The men of the French mission—six officers and six N.C.O.s—taught the Japanese how to organize, train and command units from the size of the company on up through the brigade. They also set up a school for N.C.O.s and established an academy for officer candidates, which opened its doors in 1875.[57] Apparently the Japanese government was satisfied with the work of the mission, since the contracts of the French officers were renewed in 1876.[58]

The Satsuma rebellion, demanding the commitment to battle of all the meager and still relatively untrained military forces the government had available, revealed a number of deficiencies in the Japanese army. It also had an unfortunate, disruptive effect on the work of the French mission. Essential services such as the organization of the reserves and the regular provision of supplies functioned badly. These were matters the French mission had hardly touched upon, possibly because their own army was deficient here, as had been demonstrated in the recent war against Prussia. In any case the French incurred some of the blame for the difficulties encountered in suppressing the Satsuma rebellion, while the ultimate success of the government after some six months of hard fighting gave the Meiji leaders a sense of confidence in their own abilities and in the efficacy of their new military institutions. One result was a belief that they were becoming capable of training their own military forces and that they no longer needed the tutelage of the French or of any other nation on an exclusive basis. With much tact and diplomacy, the Japanese government set about negotiating the termination of the French military mission. It went home in 1880.[59] To the degree that the Japanese utilized the services of any one European country for the rest of the nineteenth century, they turned to Germany.

The Satsuma rebellion only reinforced the perceptions of certain officers that the structure and functioning of the Japanese high command were inadequate. There did not yet exist any mechanism for overseeing the planning and the efficient conduct of military operations on a large scale, nor were there men trained in these matters. As early as 1873, Yamagata had set up at the Army Ministry a special office to develop plans and operations. Named the Staff Bureau the next year, it had little effect on the conduct of operations during the Satsuma rebellion. Then in late 1878 the government went a step further when Yamagata submitted a memorial to the emperor calling for the replacement of the Staff Bureau by a new organization, the General Staff Headquarters, with responsibility for all matters pertaining to the high command. The head of this new bureau would no longer be hierarchically subordinate to the Army Minister but would report directly to the emperor. Since the Army Minister did not have the right of direct access to the imperial personage, this gave to the Chief of Staff a position superior to that of the Minister. The enactment of the Yamagata memorial also signified the beginning of a split between the command and administrative functions within the army, one which would become institutionalized over the coming decades. That the newly established office of Chief of Staff was one possessing notable power and prerogatives vis-à-vis the army was evident when Yamagata resigned as Army Minister to take it up.[60]

By instituting the General Staff and endowing it with wide-reaching authority over the army, the Japanese were clearly following the Prusso-German example. In fact, the Japanese were some years ahead of the Germans, who did not free the General Staff from the official, if nominal, control of the Ministry of War until 1883. Yamagata saw it as one way of making the command structure of the army more efficient. He also believed that it would strengthen the unity and cohesion of the army as a whole by protecting it from politics. Possessed of obsessive distaste for what he considered to be the divisive effects of partisan politics on the national fabric, Yamagata was determined to shelter the army to the greatest degree possible from that supposed threat. One way to accomplish this, in his opinion, was to remove responsibility for a number of the most crucial aspects of military affairs from the Army Minister, a member of the governing cabinet and therefore subject to the workings of the political system, and to place it directly under the emperor. Since the emperor was ultimately a figurehead, this made the chiefs of the army responsible to no other authority but themselves for what could be a vitally important sector of national life. The command function was thereby removed from the political checks which might be exerted through the cabinet.

The army and the navy were meant to be further protected from the baleful consequences of partisan politics by the two ordinances Yamagata authored in May 1900. According to the terms of the ordinances in question, only generals or admirals on active service were eligible to be cabinet ministers for their particular branch of the armed forces. This did little more than restate what had long been the practice of the government, for the ministerial heads of the armed forces had practically since the beginning of the Meiji era been generals or admirals. But by stating that the officers in question had to be on active duty, the ordinances made sure that each would be under the influence of the chiefs of his particular branch and would represent and defend its views within the government. There was always a possibility, however slight, that high-ranking officers on the retired list might be more susceptible to civilian attitudes. Yamagata's intention here was no doubt to reaffirm more strongly the independence of the armed forces. They, unlike the political parties, which were gaining in importance in the life of the country, were assumed to embody and defend Japan's permanent general interests and not to promote some selfish factional goal.

Paradoxically enough, the effect of the various measures meant to isolate the armed forces from the pressures of politics and the perils of policy-making eventually had the opposite effect. If the policy decided upon by a cabinet was displeasing to either one of the services, or rather to the members of its supreme command, they had only to direct their particular minister to give up his portfolio. The cabinet would then be obliged to resign. When a new cabinet was in the process of being formed, elementary political wisdom would indicate that its leaders assure themselves beforehand that either in its composition or in its projected programs it would not be displeasing to the military. In other words, the armed forces were given an implicit veto and thereby a role, even if an essentially negative one, in the making of government policy.

So long as there was no conflict between the military and the politicians over matters of policy or over what the military considered to be their vital interests, the Yamagata ordinances presented no serious inconveniences. That was the case during most of the period between 1900 and 1930, but such differences of opinion did begin to arise in the early 1930s, and they had an unfortunate effect on Japanese political life. Always an important element in the life of the nation, the army began, in increasingly overt fashion, to advocate actions promoting the interests of the nation as it understood them against the purportedly selfish, unnational behavior of the civilian politicians. As has often been the case when a military force is drawn into or willingly enters a policy-making role, the factionalism characteristic of political life began to permeate it.

Yamagata and the other military leaders of his generation had sought to maintain the independence of the army and thereby protect its unity and integrity. They would not have been pleased at the situation which had developed by the mid-1930s. Factional strife within the military over the correct path for the country to follow had led groups of junior officers on occasion to rise against their superiors and even to assassinate them. Meant to be isolated from the arena of politics, various groups of officers had entered it with a vengeance, functioning as political factions like any other. Because the policies they promoted and ultimately imposed were misguided, the consequences turned out to be unfortunate for the army and the country.

7 *Conclusion*

As has been suggested in the foregoing essay, the introduction of European-style military techniques and institutions into Russia, the Ottoman Empire, Egypt, China, and Japan was a significant first step, probably an irreversible one, in the modernization of these countries. That was not the original intention of the men who launched the programs of military reform; they were not aiming to effect radical changes in the way their societies worked. All the reformers really wanted to do was to defend them against aggression from abroad, especially at the hands of certain European powers. One apparently rapid and expedient way for the countries in question to improve their defenses was through the adoption of some of the military modes of the Europeans, since the latter's superiority in war had been demonstrated in a number of armed confrontations. The reformers were to learn, often to their dismay, that the introduction of European forms and methods into their military establishments would sooner or later oblige their societies to undergo internal adjustments which were by no means trivial.

The consequences of Europeanizing military reform in countries of the extra-European world varied according to circumstances in each. Yet for all the differences among the five here treated, there were enough similarities among them with respect to why and how they undertook to establish European-style armed forces, as well as in the effects of the reform, to suggest commonality in the process they were all undergoing.

The most immediate impact of the Europeanizing military reform came as the governments in question found it necessary to raise much more money to support the new-style armed forces, if only because modern weaponry was expensive. The higher costs were also a result of maintaining the armies on a permanent basis, so that the troops might be inculcated with the high degree of discipline and receive the intensive training necessitated by the European way of waging war. Under the previous system, armies had generally been brought together at the outbreak of hostilities; it was assumed that between times a soldier could provide for his own upkeep in one way or another. If the required funds were to be extracted from the populace of a country, that presupposed a more systematic, stringent control over its tax revenues and a consequent increase in the effective power of its ruling authorities.

Although the realms in question were ostensibly autocracies governed through centralized institutions of state, the dominion of the rulers was often tenuous. Real power over the inhabitants, and with it control of the potential fiscal resources of the realm, was usually exercised by a multitude of local elites. Since the government at the center now pretended to assert its authority throughout the country, the assumed rights and prerogatives of those in the local elites were threatened. Until their opposition—whether it be in the form of overt rebellion or passive noncooperation—was overcome, the reforms seen to be necessary would be very difficult to carry out.

Resistance of a more immediate, even dangerous kind was raised by the old-style soldiers. Military reform along European lines presaged the obsolescence of the martial techniques in which they had been trained. Their professional identity and their monopoly of the military function, along with their significant role in society, were founded on their mastery of these techniques. Europeanization of the armed forces could thus be understood as an attack on some very real vested interests. It also portended the disruption of what the traditionalist soldiery saw to be the proper ordering of society. They therefore opposed military reform, often with ferocious determination. In the case of Russia, the Ottoman Empire, Egypt, and Japan, there were sooner or later violent confrontations between the reforming authorities and many of the old-style warriors, confrontations usually ending with the slaughter of larger numbers of the latter. If in China they made no serious effort to take arms against government-sponsored efforts at reform, this would seem to suggest just how degenerate the Chinese forces had become.

Any organs of centralized administration or fiscal control which were in place when the process of military reform was initiated were likely to be rudimentary, as was the case in Russia, or ineffective—even ramshackle—as in the Ottoman Empire and China. As for Japan and Egypt, such organs did not really exist. To meet the administrative and financial requirements of the modernized armed forces, the reforming authorities attempted to erect a more effective system to manage public affairs, one generally modeled on European practice and, at least in theory, more efficient and equitable in its exactions. Because that new system was in effect imported from abroad, it was not in accord with traditional mores and therefore caused opposition. Moreover, it did not for a long time work very well, partly because competent personnel trained to act according to European bureaucratic standards of honesty, efficiency, and impersonality were lacking.

The increased demands of the ruling authorities were not limited to taxes. Many in the populace, heretofore thought to be unworthy or incapable of rendering military service, were now enrolled in the ranks of the armed forces as some form of conscription was introduced to provide recruits in a more regular, predictable way. Even if most of the inhabitants of the countries under consideration were seldom moved to take arms against the government because of the greater severity of its demands, their attitude toward the new system of rule was usually one of surly apathy at best.

Both the Europeanized armed forces and the more elaborate machinery of state established primarily for their support required the services of people possessing talents not heretofore needed or recognized. Thus the proponents of Europeanizing military reform had to establish schools with the utilitarian goal of providing training in the necessary skills, for the existing educational institutions were usually under the sponsorship of the religious authorities and devoted to the teaching of religious matters. Granted that in China the content of Confucian learning was not religious, it was nevertheless hardly utilitarian. In Russia, the Ottoman Empire, Egypt, and China, schools for the professional education of officers to command the new-style military units were the first significant institutions for the dissemination of secular or modern learning, and for a long period of time they were often the only ones. In both Egypt and the Ottoman Empire, the reformers also soon realized that to handle the curriculum at the military schools, the students needed some kind of preliminary academic grounding. They therefore took the lead in organizing a network of secondary schools where candidates for the military academies could obtain suitable preparatory training, schools frequently run by army officers.

The men destined to be the officers of the modernized forces were often recruited from outside the traditional sources of military manpower. Because the new, European-style armed forces could thus bring people from what were the less eminent, less prestigious strata of society into positions of potential power and influence, they functioned as agents of social mobility within a fundamentally ascriptive order. By their origins, their special schooling, and their particular outlook, these officers constituted a disparate element within the elite.

Dissonance between the Europeanized soldiers and the rest of society was to some degree characteristic of all the countries under consideration, but in certain of them it had very disturbing consequences. Many officers, either through their training at the military academies or their experience on active duty, had been introduced to European concepts of administrative rationality and efficiency. A few may even have become open to the liberal values frequently propounded in European social or political writings. They were thus provided with a new intellectual framework in terms of which they might perceive and criticize objectionable features in their own society.

Some of the officers saw the deficiencies in the way the armed forces worked as resulting from the benighted nature of the current system of rule. Imbued with a sense of patriotism and public spirit, or perhaps merely ambition, as well as possessing access to the instruments of force, they were ready to try to remedy the situation. In the Ottoman Empire a large number in the Europeanized officer corps were adherents of the Young Turk movement. They were thus among the most determined opponents of the inefficient despotism of Abdulhamid II, while native-born Egyptian officers took the lead in the movement against the purportedly unpatriotic policies of the khedive Ismail and his successor. Officers in the newly modernized army of imperial China formed the core in the conspiratorial organizations aiming at the overthrow of the decaying Manchu dynasty, and they precipitated the uprising which finally brought it about. In all three cases the Europeanized army, instead of functioning to defend the country, and with it the political status quo, ended by being a major force for sedition.

It may be stretching the preceding argument a little far to see the 1825 Decembrist Rebellion in Russia as an ultimate consequence of the military reforms of Peter the Great a century before. Still, officers of the guards' regiments, having been inculcated with a modern, European outlook by their education and their experiences abroad, did seek to prevent the accession to the throne of Nicholas I. To them he represented what was especially retrograde and iniquitous in the Russian system. The argument may have been stretched even further in the case of Japan. Nevertheless, it was officers of the Westernized army who brought about the breakdown of the still-fragile, European-style political institutions of the country. The refractory officers, primarily of peasant and petit bourgeois origin, were acting in the name of ideals they took to be those of the traditional warrior elite, the samurai. The irony in the Japanese situation was that so much of the Westernization against which the officers were rebelling had been carried out by men of the samurai caste. Most of the latter had to a large degree rejected such ideals as anachronistic.

When a few determined men in the ruling elite of the countries under consideration began to introduce modern military methods within their societies, that was about all they wanted from European civilization. Like those they ruled, the indigenous reformers regarded with distaste—even abhorrence—the ethical codes, the social ideals, and the political practices of Europe, not to mention the incessant, restless dynamism of its denizens. What the reformers aspired to do was to borrow selectively. Each intended to adopt European military methods and utilize them to defend the traditional ways of his own country, its special values, and its essential institutions against outside aggressors—above all, the Europeans.

The reformers would seem to have conceived of the military techniques of the Europeans as being distinct and separable from the other elements of their culture. In so doing they may have been laboring under a misapprehension, but they were not necessarily being ingenuous or obtuse. When viewed by outsiders, the material, technical aspects of a civilization are usually more intelligible than are its nonmaterial features. Those technical, material aspects can thus be envisioned as providing models for imitation, while the artistic and intellectual accomplishments of an alien culture are often difficult to assimilate. As for its spiritual vision, that can be inaccessible indeed.

The military methods of a people may be seen as akin to their material and technical concerns. Military matters also tend to be considered as distinct from, and even antithetical to, the other elements in the common life of a society—the domain of a caste of trained specialists. These specialists are segregated from the great mass of the population, if only because they are dedicated to the study and practice of systematic violence, something untoward and even antisocial in the eyes of most people who are involved with more prosaic, pacific pursuits. For whatever reason, the military modes of the Europeans were looked upon as being essentially technical matters, something apart from the other traits of European culture. They were seen to have almost an autonomous existence, and by that token were assumed to be amenable to transfer and adoption abroad without leading to unmanageable perturbations within the host country. It was an assumption held by the reforming elite in each of the five countries under consideration, with the possible exception of Japan. It was also a major misconception.

How people prepare for and wage war, and the organizations they create for that purpose, are in fact closely related to the ways in which they deal with the other, more peaceable aspects of life in society. That was certainly true in the case of Europe. There the armed forces were the embodiment of qualities characteristic of its civilization: technicalism and functional rationality. Indeed the armed forces may have been the first secular organizations in the history of Europe to manifest those qualities so perfectly.

Originally established by the monarchical authorities in a number of European countries during the seventeenth century, the standing army was meant to assure the defense and further the aggrandizement of a given realm. But the impact of the army on internal developments within a kingdom tended to be of greater significance than anything it was likely to accomplish with respect to threats from abroad. Once the army had been brought into existence, once its organization had been made permanent and its mode of operation systematized, it constituted for the rulers a coercive instrument of unprecedented strength. Any exactions they might wish to make on their subjects, whether for men to fill the ranks of the army, for money to support it, or for any other purpose, were now difficult to resist because they had constantly at hand so splendid a means for asserting sociopolitical control.

The coercive capabilities inherent in the system of rule founded upon the standing army had at first to be exerted overtly. After a period of time, however, habits of obedience and attitudes of subordination were induced among the populace, with the result that the demands of the rulers came to be accepted almost without demur. Even though the theoretical prerogatives of a European monarch of the seventeenth and eighteenth centuries may have appeared to be relatively limited compared to those rulers in other cultures, his de facto power and authority were greater because he now possessed a permanently organized military force, one whose demands on his subjects met with little ongoing resistance. So great an increase in the real power of European rulers over their realms led to the breakdown of traditional, localized political institutions and procedures. These had taken root during the Middle Ages, fostered by the debility of the central authorities following the decay of the Roman and Carolingian Empires, but they could no longer stand firm once the monarchs were effectively armed.

When modern military methods and institutions were introduced into the countries of the extra-European world studied here, the outcome was quite similar to what had transpired in Europe. If anything, military reform in these societies was more upsetting than had been the case in Europe. The reformers may have hoped that through Europeanization of the armed forces the territorial integrity of their realms would be better defended, along with the existing sociopolitical order in which they clearly had a stake. They were to be disappointed. The needs of the modernized army led to the establishment of a supposedly more efficient apparatus of centralized administration, the institution of a system of secularized education, and with it the introduction into the ruling caste of people from outside the traditional elite strata. These changes seriously strained that sociopolitical order, weakening its resilience and eventually making it more susceptible to heterodox foreign influences. Rather than providing a defense against incursions from abroad, above all by the Europeans, the new-style army ended by serving as an agent, indeed a veritable channel, for the penetration of European ways. The ultimate results of Europeanizing military reform were likely to be little different from what would have happened if a country had been subjected to outright conquest by the peoples of Europe, namely, the imposition of many aspects of their distasteful civilization on the reluctant populace.

Even though in Europe the advent of military modernization may have created pressures on the accepted way of life, the social and intellectual wellsprings underlying the establishment of the standing army were consonant with many of the presuppositions characteristic of European culture since ancient times. Social organizations established according to principles of technicalism and functional rationality were not unfamiliar or uncongenial to Europeans. The demands occasioned by the army and its supporting institutions could, over time, become more easily accepted and, in their fashion, legitimate.

That kind of legitimacy was very difficult for analogous organizations to acquire in countries of the extra-European world. There was greater opposition, either active or passive, to the exactions occasioned by the new military system, so force frequently had to be used against the populace. The inhabitants of Russia in the eighteenth century, the Ottoman Empire during the *Tanzimat,* and Egypt in the reign of Muhammad Ali were to experience the new system of government and public administration as being more oppressive than what had been there before. As for China and Japan, the inhabitants of these countries did not find that the establishment of a new-style army represented a lessening in the fiscal and social burdens they had to bear. For all of these people, modernization, or Europeanization, as it was triggered by the needs of the reformed army turned out to be, at the very least, an unsettling, disagreeable experience.

If the breakdown in much of the traditional way of life was a concomitant of the Europeanizing of the army, that organization had an unwontedly predominant role in the new order which was taking shape. After the indigenous institutions possessing a measure of authority had been undermined and their power vitiated, the Europeanized army was often the only body in existence having some measure of social or political efficacy. A number of nonmilitary functions thus more or less automatically devolved upon its personnel. That would seem to be one explanation for the preponderance of the new-style military in the public life of the societies under study once the process of reform had been well launched.

Notes

1. Introduction: Army, State, and Society in Europe, 1400–1700

1. W. S. Haas, *The Destiny of the Mind, East and West* (New York, Macmillan Co., 1956), 82.

2. M. Hodgson, *The Venture of Islam* (Chicago: University of Chicago Press, 1974), 3:178–82.

3. O. Hintze, "Military Organization and State Organization," in *The Historical Essays of Otto Hintze,* ed. F. Gilbert (New York: Oxford University Press, 1975), 198.

4. M. Feld, *The Structure of Violence* (Beverly Hills/London: Sage Publications, 1977), 190.

5. M. Roberts, *Gustavus Adolphus: History of Sweden* (New York: Longman, 1953), 1:319–23.

2. Military Reform Under Peter the Great, His Predecessors, and His Successors

1. N. V. Riasnovsky, *A History of Russia,* 2d ed. (New York: Oxford University Press, 1969), 218.

2. R. Hellie, *Enserfment and Military Change in Muscovy* (Chicago: University of Chicago Press, 1971), 26–27.

3. Hellie, *Enserfment,* 1–18.

4. J. H. Billington, *The Icon and the Axe* (New York: Alfred A. Knopf, 1966), 61.

5. M. T. Florinsky, *Russia* (New York: Macmillan, 1953), 1:269.

6. Hellie, *Enserfment,* 78.

7. Hellie, *Enserfment,* 32.

8. R. Pipes, *Russia under the Old Regime* (London: Weidenfeld and Nicolson, 1974), 116.

9. Hellie, *Enserfment,* 29–37.

10. J. Keep, *Soldiers of the Tsar* (Oxford: Clarendon Press, 1985), 80.

11. Hellie, *Enserfment,* 162–63.

12. Ibid., 202.

13. Ibid., 177.

14. T. Esper, "Military Self-Sufficiency and Weapons Technology in Muscovite Russia," *Slavic Review* 28 (June 1969): 203.
15. Hellie, *Enserfment,* 349.
16. Ibid., 177–79.
17. Ibid., 192.
18. J. Blum, *Lord and Peasant in Russia* (Princeton, N.J.: Princeton University Press, 1961), 183.
19. Keep, *Soldiers,* 62.
20. J. Keep, "The Muscovite Elite and the Approach to Pluralism," *Slavonic and Eastern European Review* 48 (April 1970), 219.
21. Hellie, *Enserfment,* 216–18.
22. V. Klyuchevsky, *The Rise of the Romanovs,* trans. L. Archibald (London: Macmillan, 1970), 300.
23. Billington, *Icon,* 673.
24. Ibid., 672.
25. Hellie, *Enserfment,* 194–95.
26. Ibid., 226.
27. Ibid., 230.
28. Ibid., 201.
29. D. Tschizewskij, *Russian Intellectual History* (Ann Arbor, Mich.: Ardis, 1978), 141.
30. Hellie, *Enserfment,* 232.
31. Klyuchevsky, *Rise,* 321.
32. Billington, *Icon,* 113.
33. Florinsky, *Russia,* 1:315.
34. Klyuchevsky, *Peter the Great,* trans. L. Archibald (New York: Vintage, 1958), 11–12.
35. P. Miliukov, *History of Russia,* trans. Markman (New York: Funk and Wagnalls, 1968), 1:219.
36. Keep, *Soldiers,* 100.
37. L. J. Oliva, *Russia in the Era of Peter the Great* (Englewood Cliffs, N.J.: Prentice-Hall, 1969), 56.
38. R. Hellie, "The Petrine Army," *Canadian-American Slavic Studies* (Summer 1974), 235.
39. Ibid., 247–49.
40. Klyuchevsky, *Peter,* 81.
41. Pipes, *Old Regime,* 122.
42. Klyuchevsky, *Peter,* 64.
43. Miliukov, *History,* 242.
44. Hellie, "Petrine Army," 251.
45. Pipes, *Old Regime,* 209.
46. Florinsky, *Russia,* 1:390.
47. Esper, "Military Self-Sufficiency," 206–7.
48. A. Kahan, *The Plow, the Hammer, and the Knout,* (Chicago: University of Chicago Press, 1985), 100–101, 109–10.

49. Florinsky, *Russia,* 1:368.
50. Klyuchevsky, *Peter,* 218.
51. Florinsky, *Russia,* 1:374.
52. Klyuchevsky, *Peter,* 158.
53. Florinsky, *Russia,* 1:362–63.
54. B. H. Sumner, *Peter the Great and the Emergence of Russia* (London: English University Press, 1950), 160.
55. Florinsky, *Russia,* 1:363; Miliukov, *History,* 296–97.
56. Kahan, *Plow,* 319–20.
57. Florinsky, *Russia,* 1:358.
58. Sumner, *Peter the Great,* 159.
59. Miliukov, *History,* 1:310.
60. Keep, *Soldiers,* 131.
61. Florinsky, *Russia,* 1:417.
62. Miliukov, *History,* 1:305.
63. Florinsky, *Russia,* 1:417–18.
64. Ibid., 419.
65. Pipes, *Old Regime,* 125.
66. Keep, *Soldiers,* 125–26.
67. Klyuchevsky, *Peter,* 245–46.
68. G. L. Yaney, *The Systematization of Russian Government* (Urbana, Ill.: University of Illinois Press, 1973), 55.
69. Florinsky, *Russia,* 1:414.
70. Sumner, *Peter the Great,* 136–37.
71. Klyuchevsky, *Peter,* 219–20.
72. B. Pares, *A History of Russia* (New York: Alfred A. Knopf, 1958), 211.
73. Ibid., 242.
74. Keep, *Soldiers,* 143.
75. M. Beloff, "Russia," in *The European Nobility in the Eighteenth Century,* ed. A. Goodwin (New York: Harper-Torch, 1967), 180.
76. M. Raeff, *Origins of the Russian Intelligentsia* (New York: Harcourt, Brace & Co., 1966), 61–62.
77. J. Hassell, "Implementation of the Russian Table of Ranks during the 18th Century," *Slavic Review* 29 (June 1970): 289.
78. Ibid., 285.
79. Raeff, *Origins,* 33.
80. P. Dukes, *Catherine the Great and the Russian Nobility* (Cambridge: Cambridge University Press, 1967), 26–27.
81. Raeff, *Origins,* 70–71.
82. W. Pintner, "The Social Characteristics of the Early 19th-Century Bureaucracy," *Slavic Review,* 29 (Sept. 1970): 436.
83. M. Raeff, "Home, School, and Service in the Life of the 18th-Century Nobleman," *The Slavonic and East European Review* 40, no. 95 (June 1962): 305–6.
84. Hassell, "Implementation," 290.

85. D. Tschizewskij, *Intellectual History,* 163.
86. Raeff, *Origins,* 73.
87. Yaney, *Russian Government,* 57.
88. Ibid., 58.

3. The Reform of the Ottoman Army, 1750–1914

1. B. Lewis, *The Emergence of Modern Turkey,* 2d ed. (New York and London: Oxford University Press, 1968), 23.
2. A. H. Lybyer, *The Government of the Ottoman Empire in the Age of Suleiman the Magnificent* (Cambridge: Harvard University Press, 1913), 90.
3. Ibid., 100–102.
4. H. A. R. Gibb and H. Bowen, *Islamic Society and the West* (London: Oxford University Press, 1950), 1:52–53.
5. Lybyer, *Government,* 72–73.
6. N. Weissmann, *Les Janissaires* (Paris: Le Phenix, 1938), 42.
7. M. Hodgson, *The Venture of Islam* (Chicago: University of Chicago Press, 1974), 3:113; also, Gibb and Bowen, *Islamic Society,* 1:62.
8. A. M. Kazamias, *Education and the Quest for Modernity in Turkey* (London: Allen & Unwin, 1966), 25–26.
9. Gibb and Bowen, *Islamic Society,* 1:177.
10. C. Oman, *A History of the Art of War in the XVI Century* (London: Methuen, 1937), 766.
11. Hodgson, *Venture,* 3:129.
12. S. Shaw, *History of the Ottoman Empire and Modern Turkey* (Cambridge: Cambridge University Press, 1976), 1:171; also, Gibb and Bowen, *Islamic Society,* 1:177–84.
13. Lewis, *Emergence,* 23–24.
14. H. Inalcik, "The Heyday and Decline of the Ottoman Empire," in *Cambridge History of Islam* (Cambridge: Cambridge University Press, 1970), 1:343.
15. D. A. Rustow, "The Political Impact of the West, in *Cambridge History of Islam,* 1:675–77 (see note 14 above).
16. Shaw, *History,* 1:123.
17. Gibb and Bowen, *Islamic Society,* 1:65.
18. A. Levy, "Military Reform and the Problem of Centralization in the Ottoman Empire in the Eighteenth Century," *Middle Eastern Studies* 18 (July 1982):232.
19. Lewis, *Emergence,* 47–49.
20. S. Shaw, *Between Old and New* (Cambridge: Harvard University Press, 1971), 11.
21. Ibid., 86–87.
22. Ibid., 71–72.
23. Ibid., 84–85.
24. S. Shaw, "The Origins of Ottoman Military Reform: The Nizam-i Cedid Army of Sultan Selim III," *Journal of Modern History* no. 3 (September 1965):291–95.

25. Ibid., 300–303.
26. Ibid., 304.
27. Ibid.
28. Levy, "Military Reform," 242.
29. Shaw, *History,* 1:255–56.
30. H. A. Reed, "Destruction of the Janissaries by Mahmud II in June 1826" (Ph.D. diss., Princeton University, 1951), 45–46.
31. A. Levy, "The Military Policy of Sultan Mahmud II, 1808–1839" (Ph.D. diss., Harvard University, 1968), 139–42.
32. Reed, "Destruction of the Janissaries," 45.
33. Reed, "Destruction of the Janissaries," 90, 103–4.
34. A. Cevat, *État Militaire Ottoman Depuis la Fondation de l'Empire jusqu'à nos Jours,* trans. G. Macrides (Constantinople: Imprimerie du journal "la Turquie," 1882), 1:364.
35. Reed, "Destruction of the Janissaries," 162–64.
36. Ibid., 210.
37. Ibid., 221–22, 232ff.
38. Ibid., 238ff.
39. Shaw, *History,* 2:21.
40. Levy, "Military Policy," 197–98.
41. Lewis, *Emergence,* 90–92.
42. Shaw, *History,* 2:41.
43. A. Levy, "The Officer Corps of Sultan Mahmud's New Ottoman Army, 1826–1839," *International Journal of Middle East Studies* 2 (1971):22.
44. Ibid., 27–32.
45. Levy, "Military Policy," 386–87.
46. Ibid., 392.
47. Lewis, *Emergence,* 82.
48. Shaw, *History,* 2:50.
49. Lewis, *Emergence,* 90.
50. Levy, "Military Policy," 458.
51. Shaw, *History,* 2:43; also, Levy, "Military Policy," 506.
52. Shaw, *History,* 2:38–39.
53. Levy, "Military Policy," 517.
54. Reed, "Destruction of the Janissaries," 330–32.
55. Levy, "Military Policy," 453–54.
56. Levy, "Military Policy," 170–77.
57. D. A. Rustow, "Political Impact," 1:682.
58. Lewis, *Emergence,* 103.
59. Levy, "Military Reform," 227.
60. Ibid., 107.
61. E. Engelhardt, *La Turquie et le Tanzimat* (Paris: A. Cotillon, 1882), 126–27.
62. A. Ubicini, *Lettres sur la Turquie* (Paris: Dumaine, 1853), 1:464.
63. U. Heyd, "The Later Ottoman Empire in Rumelia and Anatolia," in *Cambridge History of Islam,* 1:366 (see note 14 above).

64. R. Davison, *Reform in the Ottoman Empire, 1856–1876* (Princeton: Princeton University Press, 1963), 94–95.
65. Shaw, *History,* 2:75, 85–86.
66. Engelhardt, *Turquie,* 116.
67. M. A. Griffiths, "The Reorganization of the Ottoman Army under Abdulhamid II, 1880–1897" (Ph.D. diss., University of California, Los Angeles, 1966).
68. Engelhardt, *Turquie,* 116.
69. Quoted in Ibid., 115–16.
70. F. Milligen, *La Turquie sous le Régime d'Abdul Aziz* (Paris: Librairie Internationale, 1868), 38–41.
71. N. Berkes, *The Development of Secularism in Turkey* (Montreal: McGill University Press, 1965), 105.
72. Levy, "Officer Corps," 33.
73. Ibid., 34.
74. M. A. Griffiths, "Reorganization," 15.
75. Shaw, *History,* 2:106.
76. Ibid., 106–7.
77. Ibid., 249–50.
78. L. Lamouche, *L'Organisation Militaire de l'Empire Ottoman* (Paris: L. Baudoin, 1895), 59–61.
79. I. Basgoz and H. E. Wilson, *Educational Problems in Turkey, 1920–1940* (Bloomington: Indiana University Press, 1986), 17–18.
80. Ibid., 18.
81. S. Mardin, *The Genesis of Young Ottoman Thought* (Princeton, N.J.: Princeton University Press, 1962), 214–15.
82. Berkes, *Secularism,* 111.
83. Shaw, *History,* 2:155–56.
84. Ibid., 97–98.
85. Heyd, in *Cambridge History of Islam,* 1:368 (see note 14 above).
86. Lewis, *Emergence,* 131–32.
87. E. E. Ramsaur, *The Young Turks* (Princeton: Princeton University Press, 1957), 5.
88. Davison, *Reform,* 331–35.
89. Shaw, *History,* 2:212.
90. Mardin, *Genesis,* 213–14.
91. Davison, *Reform,* 266.
92. Griffiths, "Reorganization," 111–13.
93. Shaw, *History,* 2:245–46.
94. Shaw, *History,* 2:225–26; also, Griffiths, "Reorganization," 85–88, 133–35.
95. Griffiths, "Reorganization," 87.
96. Shaw, *History,* 2:255–56.
97. Ramsaur, *Young Turks,* 17–18.
98. Griffiths, "Reorganization," 36.
99. Shaw, *History,* 2:263–64.

100. Ramsaur, *Young Turks,* 95.
101. Lewis, *Emergence,* 206.
102. Ramsaur, *Young Turks,* 114–15.
103. Hodgson, *Venture,* 256.
104. Ramsaur, *Young Turks,* 132.
105. F. Ahmad, *The Young Turks* (London: Oxford University Press, 1969), 163.
106. Heyd, in *Cambridge History of Islam,* 1:372 (see note 14 above).
107. Lewis, *Emergence,* 206.
108. Ramsaur, *Young Turks,* 120.
109. Shaw, *History,* 2:286–87.
110. Griffiths, "Reorganization," 3.
111. Shaw, *History,* 2:291.

4. The Egyptian Army of Muhammad Ali and His Successors

1. K. Karpat, "Transformation of the Ottoman State, 1789–1908," *International Journal of Middle East Studies* 5 (1972):251–52.
2. A. Abdel-Malek, *Egypt: Military Society,* trans. C. L. Markmann (New York: Vintage, 1968), 257.
3. D. Ayalon, *Gunpowder and Firearms in the Mamluk Kingdom* (London; Valentine Mitchel, 1956), 61–66, 76, 97.
4. H. A. R. Gibb and H. Bowen, *Islamic Society and the West,* vol. I (London: Oxford University Press, 1950), 1:227.
5. H. A. B. Rivlin, *The Agricultural Policy of Muhammad Ali* (Cambridge: Harvard University Press, 1961), 3.
6. P. Vatikiotis, *Modern History of Egypt* (New York: Praeger, 1969), 30–31.
7. P. M. Holt, "The Later Ottoman Empire in Egypt and the Fertile Crescent," *Cambridge History of Islam* (Cambridge: Cambridge University Press, 1970), 1:381.
8. A. L. Al-Sayyid Marsot, *Egypt in the Reign of Muhammad Ali* (London: Cambridge University Press, 1984), 26.
9. J. J. Richmond, *Egypt, 1798–1952* (New York: Columbia University Press, 1977), 38.
10. Cameron, *Egypt in the Nineteenth Century* (London: Smith, Elder, 1898), 82.
11. Gibb and Bowen, *Islamic Society,* 1:259–60.
12. R. Owen, *The Middle East in the World Economy, 1800–1914* (London: Methuen, 1981), 65.
13. G. Baer, *A History of Landownership in Modern Egypt* (London: Oxford University Press, 1962), 13.
14. Owen, *World Economy,* 66.
15. N. Safran, *Egypt in Search of a Political Community* (Cambridge: Harvard University Press, 1961), 32.
16. J. Heyworth-Dunne, *An Introduction to the History of Education in Egypt* (London: Luzac, 1938), 103.
17. G. Guémard, *Les Réformes en Égypte (1760–1848)* (Cairo: P. Barbey, 1936), 98.

18. H. Dodwell, *The Founder of Modern Egypt* (Cambridge: Cambridge University Press, 1931), 64–65.

19. A. L. Al-Sayyid Marsot, "Muhammad Ali's Internal Politics," in *L'Égypte en XIXe Siècle* (Paris: Centre National de la Recherche Scientifique, 1982), 161.

20. Rivlin, *Agricultural Policy,* 201–2.

21. Dodwell, *Founder,* 226–27.

22. N. Tomiche, "La hiérarchie sociale en Égypte a l'époque de Muhammad Ali," in *Political and Social Change in Modern Egypt,* ed. P. M. Holt (London: Oxford University Press, 1968), 262.

23. Rivlin, *Agricultural Policy,* 350.

24. Ibid., 349–50.

25. E. Beeri, *Army Officers in Arab Politics and Society* (London: Praeger, 1970), 306.

26. Heyworth-Dunne, *History of Education,* 114.

27. G. Baer, *Studies in the Social History of Modern Egypt* (Chicago: University of Chicago Press, 1969), 220–21.

28. Beeri, *Army Officer,* 307.

29. Ibid.

30. Marsot, *Egypt,* 127.

31. Guémard, *Réformes,* 157.

32. Marsot, *Egypt,* 156.

33. Guémard, *Réformes,* 170.

34. Ibid., 174–76.

35. Rivlin, *Agricultural Policy,* 210–11.

36. Vatikiotis, *History,* 65–67.

37. Guémard, *Reformes,* 258.

38. Rivlin, *Agricultural Policy,* 105–10.

39. Heyworth-Dunne, *History of Education,* 107.

40. Dodwell, *Founder,* 238.

41. Heyworth-Dunne, *History of Education,* 203.

42. Ibid., 104–5.

343. Vatikiotis, *History,* 463.

44. Guémard, *Réformes,* 290.

45. Ibid., 295.

46. M. Rifaat Bey, *The Awakening of Modern Egypt* (London: Longmans, Green, 1947), 40.

47. Heyworth-Dunne, *History of Education,* 286.

48. A. Schölch, *Egypt for the Egyptians: The Socio-Political Crises in Egypt, 1878–1882* (London: Ithaca Press, 1981), 23.

49. Ibid., 225–30.

50. M. Berger, *Social Change: Egypt Since Napoleon* (Princeton: Woodrow Wilson School, 1960), 10.

51. Heyworth-Dunne, *History of Education,* 227–28.

52. Quoted in Ibid., 230.

53. Ibid., 230–34.

54. Vatikiotis, *History,* 103.

55. A. Sammarco, *Les règnes de Abbas, de Said, et d'Ismail* (Rome: Instituto Poligrafico della Stato, 1935), 292.
56. Heyworth-Dunne, *History of Education*, 381–82.
57. Safran, *Political Community*, 33.
58. Marsot, *Egypt*, 165–66.
59. Owen, *World Economy*, 71.
60. Abdel-Malek, *Egypt: Military Society*, 31.
61. Rivlin, *Agricultural Policy*, 195–96.
62. Cameron, *Egypt*, 126–27.
63. Rivlin, *Agricultural Policy*, 198.
64. Rifaat Bey, *Awakening*, 46.
65. Abdel-Malek, *Egypt: Military Society*, 122–23.
66. Berger, *Social Change*, 6.
67. Vatikiotis, *History*, 73.
68. Marsot, *Egypt*, 262.
69. Schölch, *Egypt for the Egyptians*, 37.
70. Rifaat Bey, *Awakening*, 96.
71. J. M. Ahmed, *Intellectual Origins of Egyptian Nationalism* (London: Oxford University Press, 1960), 24.
72. Discussion of the title *khedive* in K. Beeri, *Army Officers*, 309.
73. Ibid., 309.
74. G. Kirk, "The Role of the Military in Society and Government: Egypt," in *The Military in the Middle East*, ed. S. N. Fisher (Columbus: Ohio State University Press, 1963), 72.
75. Schölch, *Egypt for the Egyptians*, 136.
76. W. S. Blunt, *Secret History of the English Occupation of Egypt* (London: Fisher Unwin, 1907), 130.
77. Schölch, *Egypt for the Egyptians*, 144.
78. Berger, *Social Change*, 15.
79. Ahmed, *Intellectual Origins*, 25.
80. Vatikiotis, *History*, 150–51.
81. Schölch, *Egypt for the Egyptians*, 178.
82. F. Charles-Roux, *L'Egypte de 1801 a 1882*, vol. 6 of *Histoire de la nation égyptienne*, ed. G. Hanotaux (Paris: Plon, 1936), 82.

5. The Reformation of the Armies of China, 1850–1912

1. J. K. Fairbank, ed., "Introduction," *Chinese Ways in Warfare* (Cambridge: Harvard University Press, 1974), 2. Also, L. W. Pye, *Warlord Politics* (New York: Praeger, 1971), 3.
2. W. L. Bales, *Tso Tsung-tang, Soldier and Statesman of Old China* (Shanghai: Kelly and Walsh, 1937), 50.
3. Fairbank, "Introduction," 7.
4. Ibid., 9.
5. R. L. Powell, *The Rise of Chinese Military Power* (Princeton, N.J.: Princeton University Press, 1955), 8–9.
6. F. Michael, "Regionalism in 19th-Century China," introduction to S.

Spector, *Li Hung-chang and the Huai Army* (Seattle: University of Washington Press, 1964), xxiii.

7. V. Purcell, *The Boxer Uprising* (Cambridge: Cambridge University Press, 1963), 20–21.

8. W. J. Hail, *Tseng Kuo-fan and the Taiping Rebellion* (New Haven, Conn.: Yale University Press, 1927), 8.

9. Powell, *Rise,* 19.

10. P. Kuhn, *Rebellion and Its Enemies in Late Imperial China* (Cambridge: Harvard University Press, 1970), 136.

11. F. Michael, *The Taiping Rebellion* (Seattle: University of Washington Press, 1966), 97.

12. Kuhn, *Rebellion,* 145.

13. M. C. Wright, *The Last Stand of Chinese Conservatism* (Stanford: Stanford University Press, 1957), 199.

14. Kuhn, *Rebellion,* 148.

15. F. Michael and G. E. Taylor, *The Far East in the Modern World* (London: Methuen, 1956), 187.

16. Spector, *Li Hung-chang,* 14–15 (*see* note 6, above).

17. I. C. Y. Hsu, *The Rise of Modern China* (New York: Oxford University Press, 1975), 300.

18. J. Porter, *Tseng Kuo-fan's Private Bureaucracy* (Berkeley and Los Angeles: University of California Press, 1972), 111.

19. Wright, *Last Stand,* 206.

20. F. Wakeman, *The Fall of Imperial China* (New York: Free Press, 1975), 170–71.

21. Powell, *Rise,* 35–36.

22. J. K. Fairbank, E. O. Reischauer, and A. M. Craig, *East Asia: The Modern Transformation* (Boston: Houghton Mifflin, 1965), 351–52.

23. J. K. Fairbank and S. Y. Teng, eds., *China's Response to the West* (Cambridge: Harvard University Press, 1954), 28–30.

24. Hsu, *Rise of Modern China,* 246–47.

25. T. Y. Kuo, "Self-Strengthening: The Pursuit of Western Technology," in *The Cambridge History of China,* ed. J. K. Fairbank (Cambridge: Cambridge University Press, 1978), 10:491.

26. Fairbank and Teng, *China's Response,* 50–54.

27. Wright, *Last Stand,* 210–11.

28. Kuo, "Self-Strengthening," 10:492–96.

29. Ibid.

30. Fairbank, Reischauer, and Craig, *East Asia,* 319.

31. K. Biggerstaff, *The Earliest Modern Government Schools in China* (Ithaca, N.Y.: Cornell University Press, 1961), 166.

32. Ibid., 163–76.

33. Hsu, *Rise of Modern China,* 346.

34. Spector, *Li Hung-chang,* 160–63 (*see* note 6, above).

35. Kuo, "Self-Strengthening," 10:521–24.

36. Biggerstaff, *Government Schools,* 78–79.

37. Hsu, *Rise of Modern China*, 304.

38. Spector, *Li Hung-chang*, 128–30, 151.

39. K. C. Liu, "Li Hung-chang in Chihli: The Emergence of Policy, 1870–1875," in *Approaches to Modern Chinese History*, ed. A. Feuerwerker and M. C. Wright (Berkeley and Los Angeles: University of California Press, 1967), 71.

40. Spector, *Li Hung-chang*, 169.

41. Liu, "Li Hung-chang," 77–80.

42. Wakeman, *Fall*, 193.

43. Spector, *Li Hung-chang*, 315.

44. Ibid., 275–76.

45. Powell, *Rise*, 40–41.

46. Biggerstaff, *Government Schools*, 85–86.

47. Ibid., 61–63.

48. T. L. Kennedy, "Chang Chih-Tung and the Struggle for Strategic Industrialization: The Establishment of the Hanyan Arsenal, 1884–1895," *Harvard Journal of Asiatic Studies* 33 (1973): 156–58.

49. Fairbank, Reischauer, and Craig, *East Asia*, 380.

50. Powell, *Rise*, 41–42.

51. Wright, *Last Stand*, 214.

52. Ibid., 201.

53. Quited in Hsu, *Rise of Modern China*, 341.

54. Wright, *Last Stand*, 220.

55. Powell, *Rise*, 47.

56. M. C. Cameron, *The Reform Movement in China, 1898–1912* (Stanford: Stanford University Press, 1931), 17.

57. Spector, *Li Hung-chang*, 261–62.

58. Hsu, *Rise of Modern China*, 442–53.

59. Cameron, *Reform*, 43.

60. Powell, *Rise*, 131–32.

61. Ibid., 62.

62. Ibid., 70.

63. S. R. MacKinnon, "The Peiyang Army, Yuan Shih-kai and the Origin of Chinese Warlordism," *Journal of Asian Studies* 32 (May 1973): 406.

64. J. Chen, *Yuan Shih-kai, 1859–1916* (Stanford: Stanford University Press, 1961), 79.

65. Ibid., 59.

66. Powell, *Rise*, 212.

67. Ibid., 198.

68. Cameron, *Reform*, 89–90.

69. Powell, *Rise*, 223.

70. Ibid., 258–59.

71. Ibid., 264.

72. Wakeman, *Fall*, 231.

73. M. C. Wright, "Introduction," in *China in Revolution: The First Phase, 1900–1913*, ed. M. C. Wright (New Haven, Conn.: Yale University Press, 1968), 27.

74. Powell, *Rise,* 144.
75. Ibid., 145.
76. Chen, *Yuan Shih-kai,* 84.
77. Powell, *Rise,* 230–31.
78. Ibid., 298–99.
79. Ibid., 236.
80. Cameron, *Reform,* 96.
81. E. P. Young, *The Presidency of Yuan Shih-kai* (Ann Arbor, Mich.: University of Michigan Press, 1977), 30–31.
82. J. W. Esherick, *Reform and Revolution in China* (Berkeley and Los Angeles: University of California Press, 1976), 146–49.
83. Ibid., 170.
84. Powell, *Rise,* 270–71.
85. Chen, *Yuan Shih-kai,* 115.
86. Powell, *Rise,* 80.

6. The Armed Forces of Japan during the Meiji Restoration and After

1. C. Yanaga, *Japan since Perry* (New York: McGraw-Hill, 1949), 20–21.
2. H. Kublin "The 'Modern' Army of Early Meiji Japan," *Far Eastern Quarterly* 9, no. 1 (November 1949): 23–24.
3. J. W. Hall, "The Nature of Traditional Society: Japan," in *Political Modernization in Japan and Turkey,* ed. D. A. Rustow and R. A. Ward (Princeton, N.J.: Princeton University Press, 1964), 37.
4. J. K. Fairbank, E. Reischauer, and A. Craig, *East Asia: The Modern Transformation* (Boston: Houghton Mifflin, 1965), 198–99.
5. R. Hackett, "The Military: Japan," in Rustow and Ward, *Political Modernization,* 237 (*see* note 3, above).
6. W. G. Beasley, *Modern History of Japan* (New York: Praeger, 1963), 53–55.
7. Yanaga, *Japan since Perry,* 21.
8. Beasley, *Modern History,* 51.
9. Ibid., 70.
10. Ibid., 78.
11. Ibid., 55.
12. M. Medzini, *French Policy in Japan during the Closing Years of the Tokugawa Regime* (Cambridge: Harvard University Press, 1971), 127–32.
13. Fairbank, Reischauer, and Craig, *East Asia,* 212. Also, W. G. Beasley, *The Meiji Restoration* (Stanford: Stanford University Press, 1972), 69.
14. A. Craig, *Choshu in the Meiji Restoration* (Cambridge: Harvard University Press, 1961), 21–22.
15. Ibid., 200.
16. Ibid., 232.
17. Beasley, *Meiji,* 199–200.
18. Craig, *Choshu,* 131–32.
19. Ibid., 271. Also R. Hackett, *Yamagata Aritomo in the Rise of Modern Japan, 1838–1922* (Cambridge: Harvard University Press, 1971), 39.

20. Hackett, *Yamagata,* 39 (*see* note 19, above).
21. Craig, *Choshu,* 272–75.
22. Ibid., 277–80.
23. Ibid., 312–14.
24. Medzini, *French Policy,* 134–37.
25. Craig, *Choshu,* 348.
26. Hackett, *Yamagata,* 47–48.
27. Beasley, *Meiji,* 301.
28. Ibid., 422.
29. Hackett, *Yamagata,* 57.
30. R. Wilson, *Genesis of the Meiji Government in Japan, 1868–1871* (Berkeley and Los Angeles: University of California Press, 1957), 90–91.
31. Ibid., 93.
32. Ibid., 94.
33. Fukishima Shingo, "The Building of a National Army," in *The Modernization of Japan,* ed. Tobata Seinshi (Tokyo: Institute of Asian Economic Affairs, 1966), 191.
34. Kublin, "The 'Modern' Army," 29.
35. Ibid., 52.
36. Hackett, *Yamagata,* 63–64 (*see* note 19, above).
37. Quoted in Fukishima, "National Army," 194.
38. Hackett, *Yamagata,* 67.
39. Shibusawa Keizo, *Japanese Life and Culture in the Meiji Era,* trans. C. S. Terry (Tokyo: Obunsha, 1968), 5:394–405.
40. E. H. Norman, *Origins of the Modern Japanese State* (New York: Pantheon, 1975), 394.
41. Hackett, *Yamagata,* 77. Also Beasley, *Modern History,* 118.
42. Yanaga, *Japan since Perry,* 68.
43. D. Brown, *Nationalism in Japan* (Berkeley and Los Angeles: University of California Press, 1955), 97.
44. W. W. Lockwood, *The Economic Development of Japan* (Princeton, N.J.: Princeton University Press, 1954), 17.
45. E. Reischauer, *Japan Past and Present,* 2d ed. (New York: Alfred A. Knopf, 1963), 128–29.
46. Hackett, "The Meiji Leaders and Modernization: The Case of Yamagata Aritomo," in *Changing Japanese Attitudes toward Modernization,* ed. M. Jansen (Princeton, N.J.: Princeton University Press, 1965), 257–58.
47. Shibusawa, *Japanese Life,* 5:5–6.
48. Ibid., 306.
49. Nobutaka, "War and Modernization," in *Political Development in Modern Japan,* ed. R. E. Ward (Princeton, N.J.: Princeton University Press, 1968), 196.
50. Brown, *Nationalism,* 49.
51. Kobayashi Ushiburo, *Military Industries of Japan* (New York: Oxford University Press, 1922), 161.
52. Shibusawa, *Japanese Life,* 273–75.
53. Norman, *Origins,* 227.

54. Ibid., 234.

55. Shibusawa, *Japanese Life*, 332.

56. E. Presseisen, *Before Aggression* (Tucson: University of Arizona Press, 1965), 33–40.

57. Ibid., 46–47, 60.

58. Ibid., 54.

59. Ibid., 57–59.

60. Hackett, *Yamagata*, 81–82.

Index